The Sea:

Encounter, Disencounter and Reunion

The Sea : Encounter, Disencounter and Reencounter

Education, Volume 1

Artur Victoria and Almirante Antonio Silva Ribeiro

Published by Edições Esgotadas, 2024.

While every precaution has been taken in the preparation of this book, the publisher assumes no responsibility for errors or omissions, or for damages resulting from the use of the information contained herein.

THE SEA : ENCOUNTER, DISENCOUNTER AND REENCOUNTER

First edition. November 28, 2024.

ISBN: 979-8230311539

Written by Artur Victoria and Almirante Antonio Silva Ribeiro.

2024 Artur Victoria.

©All rights reserved.

Book Cover by Artur Victoria

First edition 2024

INDEX

1. - Port Security
2. - Security in Port Traffic
3. - Logistics Platforms 5: Shipbuilding and repair

5. - Hydrographical surveys 12 - Maritime Services
1. - Shipping
2. - Port services

1. - Maritime logistics
2. - Security
3. - Technology
4. - Insurance and Finance
5. - Education and training
6. - Brokers
    1. - Ship brokers
    2. - Charter brokers
    3. - Freight Brokers
    4. - Port agent
    5. - Bunker brokers
    6. - Insurance brokers
    7. - Maritime legal service
    8. - Maritime lawyers
    9. - Classification societies
    10. - Organisations and associations
    11. - Consultants and inspectors
    12. - Marine surveyors
    13. - Maritime Engineering Consultants
    14. - Safety Consultants
    15. - Environmental Consultants
    16. - Security consultants
    17. - Logistics and Supply Chain Consultants
    18. - Legal and regulatory consultants
    19. - Port and Terminal Consultants

1. - Maritime Information and Communications Technology
    1. - Shipboard Communication Systems
    2. - Vessel Tracking and Monitoring
    1. - Maritime Safety and Navigation Systems
    2. - Port and Terminal Management Systems

3.   - Maritime Cyber security
4.   - Remote Monitoring and Maintenance

1. - Maritime protection and indemnity 16 - Strategic Thinking
    1. - Navigating Uncertainty
    2. - Capitalizing on Opportunities
    3. - Managing Risks
    4. - Enhancing Competitiveness
    5. - Promoting Sustainability
    6. - Fostering Collaboration
    7. - Adapting to technological innovation

1. - Maritime Environment and Nature Conservation
    1. - Habitat Conservation
    2. - Species Protection
    3. - Marine Pollution Control
    4. - Climate Change Adaptation
    5. - Sustainable fisheries management
    6. - Maritime Spatial Planning
    7. - Public Awareness and Education 18 - Exclusive Economic Zones (EEZs) 19 - Climate changes
    1. - Rising Sea Levels
    2. - Ocean Warming
    3. - Ocean Acidification
    4. - Changes in Marine Habitats
    5. - Extreme Weather Events
    6. - Melting Polar Ice Caps
    7. - Impacts on Marine Industries 20 - The Navy
    1. - The Military use of the Sea
    2. - Power Projection
    3. - Control of Sea Lanes
    4. - Sea Denial
    5. - Amphibious Operations

1. - Anti-Access/Area Denial (A2/AD)
2. - Maritime Security Operations
3. - Humanitarian Assistance and Disaster Relief (HA/DR)
   21 - The Navy

21 - Types of Navy vessels 22 - Navy Academies
Conclusion
Bibliography

Foreword

Admiral Henrique Eduardo Passaláqua de Gouveia e Melo

Chief of Navy Staff of the Portuguese Navy

It is with great pleasure that I preface MAR: Encounter, Disencounter Reunion, a work that delves deeply into Portugal's historical, cultural historical, cultural, economic and strategic interactions between Portugal and the sea.

Sea, without forgetting the crucial role of innovation and technology—with hitherto unimaginable projects—in the various maritime activities.

This book also warns of the need to consciously utilise this resource, which is essential for the survival of the planet and the world's of the world's population.

This book, the result of a partnership between Admiral António Silva Ribeiro and Dr. Artur Victoria, not only revisits the glorious past of Portugal, which ventured out onto the waves of the Atlantic.

It proposes an in-depth reflection on the challenges and opportunities that the maritime future holds.

The sea has played an undeniable role in shaping the identity of Portugal as a nation. The sea, more than just a natural frontier, was a means of expansion, economic development, and connection with the world.

Here, the reader explores the phases of Portugal's relationship with the sea - the encounter, the disagreement and the reunion -

and how each one reflected the challenges and opportunities faced by the country.

6

This book begins with a detailed analysis of Portugal's 'Encounter' with the sea. It recounts how, following the consolidation of borders and pacification with the kingdoms of León and Castile, the nation turned to the ocean, ushering in the era of discovery.

This moment was characterised by a series of visionary leaders who realised the strategic and economic potential of the sea.

These leaders helped position Portugal as a world maritime power. I like to say that Portugal was the first global power to understand the essence of the sea.

maritime power, far beyond the sea as a mere space for manoeuvre, access, encirclement, and projection of power onto land. The 'Disencounter', a phase characterized by constant conflict and the gradual decline of the overseas empire, invites the reader to

explore the complex dynamics that led to it.

Portugal, once a renowned naval power, faced a long period of adversity that culminated in the loss of its dominant position on the seas and a gradual retreat from it. This mismatch not only marked the decline of the Portuguese empire, but also the loss of the centrality of the sea in the country's economic and cultural life.

Admiral Silva Ribeiro clarifies, through a geopolitical and geostrategic analysis, the reasons why the country was unable to maintain its maritime leadership for centuries—for centuries to maintain its maritime leadership—"Portugal'did not have the opportunities or conditions to align and motivate political, economic, cultural, and security agents. Players in politics, economy, culture, and security collaborated on a project that would centre national life around the sea.

However, the story doesn't end with a disagreement. We are invited to analyse the contemporary conditions that could make a Portugal's 'reunion' with the sea. The 21st century has brought about the necessary stability for countries to once again look to the ocean as an engine of development.

Here, Dr. Artur Vitória does not limit himself to a nostalgic vision. On the contrary, he proposes a holistic and strategic approach to tackling modern maritime challenges, ranging from environmental issues, scientific research into the sea, nautical leisure activities, maritime services, and maritime transport to maritime safety.

This foreword would be incomplete if I didn't mention the contemporary relevance of relevance of this work.

As society increasingly acknowledges the oceans as frontiers for natural resources, biodiversity, and trade, we find ourselves at a pivotal juncture where we must redefine our global role.

This book delves deeply into critical and topical issues, such as the sustainable exploitation of the seas, the protection of exclusive economic zones, and technological innovation in the blue economy. It is essential to sensitise decision-makers, companies and civil society in general, society in general, to the various ways in which the use of the oceans can contribute to people's well-being.

However, it is also very important to be careful when protecting the marine environment as a whole and when looking at international maritime law to make sure that the different areas it covers are properly regulated and put into action, as well as that safety rules are followed. This is especially important when it comes to the exclusive economic zones of each nation and the coastal zones. This is to protect each country's sovereignty.

By immersing himself in this book, the reader will be invited to reflect on how the past has moulded the present and how today's choices will will define the future. The sea, as an element of connection and confrontation, is both a legacy and a challenge.

Both a heritage and a challenge, a symbol of identity and a field of unexplored opportunities.

And it is in this vast and mysterious sea that Portugal finds its true vocation.

The true vocation continues to inspire and challenge generations with new encounters, disagreements, and reunions.

I invite you, dear reader, to embark on this journey and allow your ideas about the sea to be challenged.

We want to challenge, enrich, and, above all, inspire your ideas about the sea.

The sea could become a very strong brand of Portugal, which is today still poorly structured and explored. It should incorporate Portugal's Portugal's historical relationship with the oceans and with other peoples, involving knowledge, technology, culture and economy

Due to the current relevance of my thoughts, I reaffirm that the sea provides Portugal with strategic depth, bolstering its independence, external connectivity, importance, and role within the main alliances, namely NATO, the EU, and the CPLP.

In fact, a brief historical analysis will show that Portugal's geostrategic value was - and is - deeply leveraged by the position and size of the inter-territorial maritime space.

As a result, it is important to connect a technological, industrial, commercial, and electromagnetic energy "cluster" to the sea and to use and develop this space effectively for its economic, strategic, and political purposes. This means building a "blue economy" that fits into the state's larger geo-economic and geo-strategic plans.

Today, the technologies, industry, and services related to seabed exploration, ocean aquaculture, wind-powered energy production, and maritime agitation are rapidly developing and have significant economic potential.

It is also foreseeable that in the medium-term future, humans will colonise the sea, transforming geopolitics, which is centred on large land masses, into a new oceanic land mass, creating a new oceanocentric reality.

Because of the roles it performs, the Portuguese Navy will always be the cornerstone of the blue economy (defense, security, knowledge), but the investments required for its upkeep could make it a powerful economic and technological catalyst.

Portugal is a country that was born of and for the sea. We are a Maritime Portugal

Brazilian Navy Supreme Commander Admiral Marcos Sampaio Olsen

...And if there were more worlds, they'd get there They would fear the sea more than the land But the famous garrison of Pedro Álvares Where the land ends and the sea begins...

Evoking the poet Luis Vaz de Cames reminds us of intrepid sailors who set sail, guided by the audacity and science of the seas, to explore new continents and expand horizons that set the course for civilisations.

The sea, which was once a fear and a challenge, has become a vital link between nations.

Brazil and Portugal, with inseparable ties that forged the waters of the Atlantic Ocean and shared the same language, share historical and cultural legacies rooted in the maritime epics that have moulded both peoples.

Browsing through the pages of this distinguished work, the reader is instigated to and multifaceted examination of Portugal's relationship with the ocean.

By means of a detailed and comprehensive analysis, the authors take a journey that covers everything from the beginnings of Portuguese naval exploration to the contemporary challenges facing a society dedicated to 'Things of the Sea'.

The initiative revisits the moments of apogee and crisis that marked Portuguese maritime history. At the same time, it offers a re-encounter with the sea, emphasising the need for a holistic vision that places the ocean at the epicentre of the country's political and social agenda.

11

For the Brazilian state, the sea transcends the condition of a mere natural frontier; it is a structuring element of the country's geopolitical geopolitical identity and is a source of opportunities, challenges and responsibilities that require preparation, strategic vision and leadership.

Emphasised by introspective analysis, the book does not limit itself to celebrating past achievements, but calls for deep meditation on Brazil's future Brazil's future prospects as a maritime power. The text emphasises the urgent need for continued investment in the Navy and in the development of a maritime mentality that goes beyond the limits of military institutions and spreads far and wide in society.

The importance of naval academies is also evident, which, over the centuries, have trained military personnel, authentic leaders, and defenders of maritime heritage who are capable of facing contemporary challenges and exploiting the resources that the sea offers.

Simultaneously, the Navy plays an educational and inspirational role, instilling in men and women values such as discipline, dedication, and loyalty, preparing them to serve as both "sentinels of the sentinels of the seas" and motivators of sustainable maritime development.

The purpose of this preface is to emphasise the importance of this distinguished compendium, which looks at the past and casts an attentive eye on the future.

Above all, it emphasises that the sea is an essential element of national strategy—integrating security, progress, and sustainability.

We invite the reader to delve into these pages and contemplate the historical legacy and potential of Brazil and Portugal as maritime nations, ensuring that the sea remains a symbol of identity, strength, and prosperity for future generations.

All for Brazil and the Navy!

Dean of Universidade de Coimbra. Professor P.h.d. Amílcar Falcão

The sea was at the origin of life as we know it. Going back billions of years ago, the planet was not as it is today and the sea itself (oceans), in its early days, didn't have the same geographical composition or geographical distribution.

The timeline made its adjustments until we reached the present day (i.e. the most recent millennia).

What is certain is that to talk about the sea is to talk about life. All land animals came from a very distant relative that one day came out of the sea. We're talking about tremendous time scales, but we shouldn't forget them. it's important not to forget them.

Human beings' attraction to the sea is very curious. as if it were a return to our origins. We like to interact with the sea in the most varied ways. The sea fascinates and respects us.

Biologically, we are very close to the sea. The human body is about 70 percent water (varying slightly with age). Interestingly, a significant portion of the water in our bodies closely resembles the composition of seawater.

Anyone who has had the opportunity to savour their tears or your sweat, you've certainly realised that what we feel is very close to (salty) water.

It's not strange, you realise, but it's striking and reinforces how our body maintains a physicochemical very close to its origins.

Human beings can (over)live for a long time without eating. There are records of shipwrecked people who have lasted 50 days without food. However, nobody can last more than 4-5 days without

13

drinking water. Physiologically, our our dependence on water is enormous.

In fact, in certain circumstances, it's not enough just to drink water (which always has some degree of mineralisation). For instance, when an athlete, such as a cyclist, loses 4 or 5 kg during a competition (essentially water), drinking water is not sufficient for rehydration.

Important quantities of mineral salts are lost with sweat. Minerals are lost, which is why water replacement must be supplemented with an appropriate amount of these same electrolytes. This is the only way the body regains a healthy chemical balance.

Because water is life, it's not surprising that when we study universal history, we easily realize that an important part of conflicts originated (directly or indirectly) in access to water.

Human life cannot be isolated from its ecosystem. That's why in addition to our umbilical connection with the sea, there are many other other interactions that cannot be ignored. This book's text addresses many aspects of this interaction.

Depending on the reader's sensitivity, there will be chapters that are more or less gripping. It happened to me when I read the book, as will certainly happen to everyone who reads it. This is a natural outcome, as it is influenced by our background, concerns, interests, and curiosity.

But that's also why that this book is capable of reaching very different audiences. And that is in literary and even emotional terms. I've already mentioned something that touches me personally when we talk about the the origin and evolution of life.

My academics forced me to start with the origin of things. However, I can't help but touch on a few other perspectives that according to my sensitivity and interest.

When I'm asked, as an academic, what I see in the sea, the beginning of my answer is invariably the beginning of my answer is invariably the same: I like to think about what the sea can give us, and what we can give the sea. In fact, this conversation starter can be summed up as the sustainability that we must seek to guarantee so that future generations can enjoy the sea (and the planet) as previous generations did.

At the beginning of time, the sea gave us food. The sea gave us mobility. During the first phase, navigation evolved into the discovery of routes that connected continents, stimulating trade and allowing the crossroads of peoples and cultures.

Fierce naval battles took place during the great wars (and not just those in the first half of the 20th century).

In the second half of the last century, the sea became very popular for leisure and rest periods.

More recently, the sea has also become a source of clean energy and minerals. All this is part of my first observation, which is "what the sea can give us".

The problem is that, as we approach the present day, we've also turned the sea into a "dustbin". I'll explore further this theme, but it's very obvious that we've been neglecting the sea's "health" .

It turns out that at a time when the concept of "one health" is concept is gaining momentum, our health also depends on the health of the environment and, consequently, marine health.

Climate change is humanity's greatest challenge. Delaying actions or focusing on other matters will have significant consequences in the future.

Anyone who argues that climate change has always existed is absolutely right. The mistake lies in assessing the speed at which they are occurring.

The phenomenon of global warming has caused an average sea level rise of between 15 and 25 cm between 1901 and 2018, according to hard data.

However, if we focus only on the 21st century, we see that the global average sea level rose by an average of 2.9 mm/year between 2001-2010, a rate that almost doubled to 4.5 mm/year between 2011 and 2020.

The immediate consequences of seawater warming include the expansion of the water mass (increase in volume) and the melting of glaciers (further increase in volume), which pose a short- to medium-term threat to all coastal locations.

Unfortunately, the problem doesn't stop there. The increasingly warmer and more acidic waters negatively affect the balance of marine ecosystems. Damage to coral reefs and a reduction in the diversity of marine species are just a few examples of the negative effects of climate change.

All these changes together alter sea currents and lead to more extreme weather phenomena, which in turn affect fish migration.

The change in fish migration means that species that we were used to fishing off our coasts are replaced by others that, due to all these changes (climate, temperature, currents...), end up relocating in a fight for survival.

The impact of these changes on ecosystems and even on eating habits is obvious. What is not obvious is the scale of this impact. Predators can become prey (and vice versa) due to the alteration of food chains, and human intervention cannot significantly alter this situation once the chaos has subsided.

So when some "optimists" say that an increase of between 1.5-2 degrees Celsius in the average temperature, it doesn't even seem that much. It may not even seem like much, but it's more than enough to cause irreversible damage to the planet.

To not be environmentally aware is to ride the wave of negativity.

When I referred above to the fact that the sea has become a "dustbin", I was trying to describe a real (and dramatic) scenario. Among many other examples, let's just think of plastics (forgetting heavy metals, oil spills, sewage discharges, etc.).

The United Nations Environment Program (UNEP) warns that the oceans receive around 13 million tons of plastic every year, the equivalent of dumping a truckload of garbage into the ocean every minute (that's terrifying...). The degradation of plastic in the sea takes centuries.

In the meantime, during this process, we have smaller plastics, reaching smaller plastics, reaching nanoplastics (particles invisible to the naked eye). Nanoplastics enter the food chain and work their way up until they reach human beings. We should refer to a recent scientific study that detected nanoplastics in breast milk to assess the seriousness of the situation. Breast milk that from the mother's blood...

Can we avoid this whole catastrophe? Probably not, but we must act to prevent the situation from getting worse.

Many people (the scientific community and society in general) argue that technology will help solve the emergency situation we are experiencing.

I personally also believe that technology will play its part in mitigating the effects of climate change, but it is clearly counting on the uncertain.

At the time of writing, there is the idea that artificial intelligence (AI) is there to solve all of our problems, including those resulting from climate change.

Although I'm not an expert in AI, I have sufficient knowledge of computer science, mathematics and statistics to understand where we are.

Not long ago, I was in a working session with a colleague, an AI expert We both gave our speeches, and, because we were personal friends, I privately asked him if he truly believed what he had said at the end.His first reaction was to let out a laugh. I persisted and defended my questioning of him.

The answer was disconcerting: we're not in the age of AI, real AI won't be around for a few decades, let's hope the world gets there and that, in the meantime, find a form of energy capable of supporting AI.

I think it's useful to clarify that the way in which mathematical and statistical models are applied today mathematical and statistical models are applied to a wide variety of situations, are based in many cases on theoretical concepts developed many decades ago (Einstein postulated laws that still need to be proven today - time/space travel is one of the most time/space is one of the best known).

Therefore, many of the technological limitations that we have experienced for decades were not related to the lack of imagination or creativity of brilliant scientists but rather to the difficulty in proving these ideas. Perhaps the biggest bottleneck in technological evolution has been the difficulty in having hardware and software that human imagination.

It's worth bearing in mind that the level of computing used by NASA to put humans on the moon is something unimaginably simple today.

Any cell phone can process (in quantity and speed) thousands/ millions of times (depending on the model) what the Apollo 11 computer could do.

Anyone who works in the field knows that today's AI is nothing more than processing more information in much less time. The machine writes the algorithms and performs a task that would take an infinite amount of time to complete by hand.

This means that AI (let's call it that for sympathy's sake) doesn't have enormous potential to help us mitigate some of the problems resulting from climate change.

It certainly does.

The combination of sensors, robotics, georeferencing, real-time images, etc. can be used to monitor many problems, anticipating some and problems, anticipating some and eliminating others.

We currently have real-time maps of seismic phenomena. We have real-time maps of the water temperature in the different oceans. We have the capacity to monitor the migration of species in real time (marine and other). We have the capacity to detect fires, preventing large-scale natural disasters. We even have the ability to monitor space debris, preventing announced disasters (yes, space has also become a very reasonable "dustbin").

Therefore, we must be aware of the situation and take all necessary steps to resume sustainable habits, which, at the very least, prevent a rapid deterioration of our current state.

Because I'm not a negative person, nor do I want to send out a negative message, I'd also like to look at the bright side of things.

We can start with a sunset by the sea in good company. What could be more relaxing? How about we talk a bit about activities that require the sea?

I've been on a boat before, but let's just say I wasn't particularly fascinated. However, there are those who have this fascination, which is respectable and justifiable. There are numerous other activities available, such as bathing in the sea, underwater fishing, and surfing.

Having personally experienced any of these activities, I can confidently state that a refreshing swim in the sea always rejuvenates. If the sea is calm, relaxation comes from the context. If the sea is rough, relaxation comes from tiredness. In either case, the sensation is always very pleasant.

As for underwater fishing, it's a risky activity, but one that unique moments. The underwater beauty is indescribable.

Whether it's due to the marine animals or the corals and algae where they live or hide.

Live or hide. The moments of solitude you get are absolutely unique and would never be possible without the sea.

As for surfing, the challenge of catching a wave is inexplicable. The process of counting waves, positioning correctly, and maintaining balance is all captivating. Now, believe it or not, there's nothing that frees our minds more our minds more than waiting for a wave and enjoying the moment.

I just don't recommend it, because surfing an easy sport and involves reasonable risks.

Then there are delicious moments like this one that I'm going to share. A friend of mine once told me in conversation that, every now and then, he likes to go fishing (with a rod). Curious, I asked him how his fishing usually went. His answer was: very well! Faced with this answer, I asked him more or less how many fish he usually brought home after a morning of fishing. The answer is none! I couldn't help laugh and ask why he was so happy.

He explained: You know, I always take two rods. I prepare one and cast the hook. Meanwhile, I prepare the other and set the hook By the time the second one is ready, the first one already requires a bait change.

In a nutshell, you spend the morning putting bait on the hook, setting the hook, picking up the hook, and so on.

With popular wisdom, he told me like this: you know, I don't fish at all, but during those hours, believe me. I don't think about anything else, and nothing bothers me, That's what the sea is all about!

I wouldn't be at peace with my conscience if I didn't exalt the the importance of producing knowledge about the sea.

As the University of Coimbra is a centuries-old institution, there is no records of its connection to the sea, particularly the influence influence it had on Portugal's expansion overseas.

The Portuguese elite formed and gathered in Coimbra, which was also the site of the most developed national knowledge production at the time. Mathematics, astronomy, and cartography, among other subjects, were then also part of the academic community.

I can't therefore end this preface without leaving one last to refer to the (still) recent launch of the Campus of the University of Coimbra at Figueira da Foz.

With its existing historical background, having in its area of influence a gateway to the sea, as is the case with Figueira da Foz, a bathing area par excellence for thousands of citizens from the region (and beyond), with the city of Coimbra at its head, the absence of the University of Coimbra in a coastal city was a gap that urgently needed to be filled. Reading the lines I wrote earlier, you can see that my perspective on the importance of the sea is very mature.

As Rector of the University of Coimbra, I considered it strategic to deepen our relationship with the sea. And this can't be achieved with isolated projects or fleeting presences.

It was my understanding that a permanent physical presence in Figueira da Foz was a strategic objective of the utmost importance for the University of Coimbra and for the country.

Training staff, producing knowledge, and collaborating with local development (both social and business) require a structured investment.

The process is underway and will certainly take time to sink in (first you get strange and then you and only then), but I'm sure it will be a winning bet.

This book adds a lot of value to the University of Coimbra's presence in Figueira da Foz.

Upon reading it, it's clear that there was an error that needed correction.We extend our congratulations to the authors of this excellent work.

The Sea: Encounter, Disencounter and Reunion

Admiral António Silva Ribeiro

23

Introduction

This text explains the circumstances of Portugal's transformation into a Portugal's transformation into a maritime nation, its distortion and the possibilities of the sea once again become a central element of national life. It begins with the characterisation of Portugal's encounter with the sea, which took place after the consolidation of the borders of the national territory and the consummation of peace with the kingdom of León and Castile, from which the charismatic and enlightened leaders of the 14th enlightened leaderships of the 14th, 15th and 16th centuries led the country into a period of period of political stability, economic and social development, and the affirmation of the Portuguese mentality and maritime activity, paving the way for the the age of discovery.

The determinants that led to Portugal's mismatch between Portugal and the sea, resulting from a state of permanent war, which eventually led to the end of the expansion, the decline of the overseas empire, the progressive loss of status in the hierarchy of powers and the continued erosion of the national mentality and mentality and national maritime activities. : Using historical facts and contextualizing the geopolitical and geostrategic environment in which the country found itself between the 16th and 20th centuries, we can understand why Portugal was unable to align and motivate political, economic, cultural, and security actors with a project that would focus national life on the sea.

More than 800 years have passed since the beginning of Portugal's encounter with the sea.

Portugal's encounter with the sea, the 21st century has once again brought security to the forefront.

political stability, economic, and social development.

As a result of economic and social growth, there are now good conditions that could lead to a new encounter with the sea if the

24

national maritime activity can be revived and the mindset can be fixed.

In this context, it is important to recognise that the country is confronting a wide range of maritime challenges, including cultural, political, economic, environmental, and security issues.

To figure out how to deal with these problems and get back in touch with the sea at the same time, we need to use an analysis model that takes a comprehensive and strategic view.

This is necessary to leverage the charisma and enlightenment of national leaders, ensuring greater levels of development and security for future generations.

The Encounter

Portugal was born by the sea, and its formation was influenced by the geographical imperatives of its position.

However, in the first century of its existence, the country did not have a national maritime way of life, national maritime lifestyle, as the reconquest and territorial territorial consolidation directed the efforts of the nation's builders towards nation-builders to contain the Leonese and Castilians along the northern and eastern borders, as well as to north and east, as well as conquering territories in the south from the Almoravids.

The aim of protecting the Crown's assets and ensure independence and the constant pressure to which the Portuguese land borders were subjected between the 12th and 13th centuries contributed to the country's the most assertive option for affirming the country was through the sea, as it was the better possibility of permanent contact with the outside world, without obstructed by other actors.

To this end, from the sea came support from the Crusaders, a reflection of what was a brilliant strategy of our bishops, who convinced them that it was as sacred to fight the infidel in the Peninsula as sacred as liberating the Holy Sepulchre!

It was within the framework of these strategic processes that Portugal integrated three fundamental maritime assets into its heritage: Lisbon in 1147, during the reign of King Afonso Henriques; Silves in 1189, during the reign of Sancho I; and Alcácer do Sal in 1217, during the reign of King D. Afonso II. These cities, along with Porto, which gave its name and was the genesis of the national geohistorical centre, were essential to the Portuguese maritime mentality and activity in culture, politics, the economy, and security.

26

Of course these cities played another fundamental role, as they contributed to consolidating the importance of Braga, Guimarães, Coimbra, Leiria, and Santarém, decisive points in the reconquest and consolidation of our territory.

It was also in the 12th century that Portugal began to develop a national way of life, with the significant economic contribution of the maritime sector.

The first documents relating to Portugal's maritime activities date from this period. :Maritime activities demonstrate the importance of fishing and maritime trade in the Portuguese medieval economy.

The establishment of a trading post in Flanders during the last quarter of the 12th century and the subsequent strengthening of maritime trade with England in the early years of the 13th century are also noteworthy in this context.

The Treaty of Alcanizes, which defined Portugal's borders with the kingdoms of León and Castile in 1297, resolved several areas of conflict and stopped minor territorial disputes from escalating into larger-scale wars.

The agreement represented a fundamental contribution to the consolidation of the Portugal's encounter with the sea, insofar as, while previous conflicts had previous conflicts made it possible to incorporate three fundamental maritime assets, as well as fostering economic development through the maritime sector, the peace that resulted from them made it possible to exploit them in favour of the the national maritime way of life.

The consummation of these favourable circumstances for Portugal's encounter with the sea with the sea was possible because leaders with vision, courage, audacity and determination. First and foremost, D. Dinis, with his perception of wide horizons, based on a determined impetus, with personality, originality and boldness, revealing an ability to align and motivate cultural, political

Portugal's cultural, political, economic and secondary agents towards maritime maritime causes. In this context, the Navy was created in 1317 on a permanent basis.

In 1317, the creation of the Navy aimed to promote the mercantile network with Britain, England, and Flanders while also ensuring the safety of national maritime activities.

To increase clarity, replace with: The nationalisation of the Knights Templar led to the foundation of the Order of Christ in 1321, initiating a national maritime strategic mindset.

From 1361 to 1385, Portugal once again found itself involved in wars against Castile, where it had to fight to preserve its independence and the maritime assets that the country had integrated into its sovereignty. The victory of Aljubarrota in 1385 finalised the confrontations with Castile, leading the country from peace in 1411 to a period of political stability and economic and social development.

It was within this framework that Portugal's encounter with the sea was consolidated peace with the traditional enemy, allowed the recovery of the organic balance of the State, paving the way, under the Avis Dynasty, for one of the most remarkable one of the most remarkable periods in Portuguese history, due to the affirmation of the maritime mentality and maritime activity in the cultural, political, economic economic and security fields.

It's important to note, however, that the consolidation of the maritime way of life was only possible because, on the one hand, the country, at peace on the one hand, the country, in peace, focussed, in an integrated manner, all of its resources, capacities and competences on maritime projects.

In this context, the field of nautical science is of the utmost importance differentiating factor of Portugal's competitive advantage at sea.

On the other hand, and once again, the emergence of charismatic and enlightened leadership

Leadership was decisive, particularly those led by D. João I and João II, the 'Perfect Prince' of Portuguese strategic thinking, with his sublime strategic thinking, with his sublime strategic vision that, in relation to Castile, it was necessary to contain it on land and beat it at sea! Keeping Castile contained on land meant maintaining peace.

Beating it at sea meant developing of a distinctive national way of life linked to the sea, capable of of aligning and motivating Portugal's cultural, political, economic and security agents to maritime causes King João II's focused approach on a common goal created the synergies needed to bring culture and maritime affairs closer together. This approach also shifted diplomacy toward maritime issues, boosted the maritime economy, and put the naval and land forces to work on the maritime-based expansionist project.

King Manuel I initially continued the maritime actions of his predecessors, which led to the discovery of the sea route to India, Brazil, and the coveted spice islands, the Moluccas, which were crucial for the expansion of the Portuguese Empire.

Consequently, at the beginning of the century, Portugal assumed the immense responsibility of organising and administering vast overseas territories, which were spread over three continents, with portions located within the boundaries of the then known

Among them were the maritime routes that fueled the movement of people, goods, services, technology, and capital on a global scale and in progressively increasing quantities. and in progressively increasing quantities. Perhaps this marked the beginning of globalisation.

It was in this context that, in 1499, the monarch, after Vasco da Gama returned from India, reformulated the title he had inherited from King João II, 'King of Portugal and the Algarves, of Africa and beyond, and Lord of Guinea', adding 'and of the Conquest,

Navigation and Trade of Ethiopia, Arabia, Persia and India, etc.' As a result the adoption of the strategic vision intrinsic to his nobiliary title, the peaceful prosperity provided by maritime trade did not not last long, as the monarch determined a significant change in the national significant change to the national strategy, expanding it, as his title attests, to territorial conquest, which ultimately led to Portugal's mismatch with the sea and, irreversibly, the end of expansion, the decline of the overseas empire, the gradual loss of status in the hierarchy, the progressive loss of status in the hierarchy of powers and the erosion of the national maritime mentality and activities.

The Disencounter

The beginning of Portugal's disagreement with the sea occurred when the war, as a result of the aforementioned strategic change, became an important national priority to which all resources, capacities and competences had to be directed resources, capacities and competences in order to ensure military, political military, political, cultural and economic domination over territorial conquests.

In this context, we can say that the disagreement had a significant impact in 1509, shortly after the naval battle of Diu, which was undoubtedly one of the most emblematic of the Portuguese Navy. In tactical terms, it was a decisive or annihilating battle; one of the contenders, the Muslim coalition (Turks and Egyptians), lost its naval power. However, soon after this extraordinary victory, Francisco de Almeida handed the government of India over to Afonso de Albuquerque, who, with increasing belligerence, materialised a policy of territorial empire based on the conquest of maritime trade centres with the aim of closing the Indian Ocean to Muslim navigation.

This strategic course of action resulted in a perpetual state of war against the Indian and Malay sovereigns, leading to arduous and costly land campaigns and enduring Portuguese overseas possessions.

In another context, King Manuel I's determination to broaden the national national bases in literal Morocco, met with a vigorous Muslim defence of Muslim defence, capable of imposing various setbacks on Portuguese attempts

In addition, Muslim pressure pressure on our African plazas increased, rendering the ephemeral commitment and heroism of its defenders The culmination of Portugal's disagreement with the sea was, in 1578, the battle of Alcácer-Quibir, which resulted in the Portuguese defeat, with the disappearance of King Sebastião in

31

combat and the imprisonment or death of illustrious members of the Portuguese nobility.

This African battle marked the end of the Avis Dynasty and the period of maritime expansion that began with the victory of Aljubarrota. A dynastic crisis led two years later to Portugal losing its independence to Spain.

The Iberian Union, which lasted from 1580 to 1640, significantly changed Portugal's status and relations in the international system and was characterized by a period of absence of national foreign policy. Portuguese maritime determinants in international relations, territorial organisation and trade, harbours, language, customs and strategic alliances were, subjugated to Spain's visions in the cultural, political, economic and security fields.

In 1640, King João IV restored Portugal's independence. However, the country became embroiled in the drama of the War of the Restoration, from 1640 to 1669, which continued in the War of Succession, from 1701 to 1714, and the Seven Years' War, from 1756 to 1763.

A maritime vision did not control the way people lived in the country because of the short time between these conflicts and the political and military situations that required all of the country's resources, skills, and abilities to be used.

Nevertheless, the alliance with England remained at sea, which helped preserve colonies and national independence. In the economic sphere, the Marquis of Pombal established prestigious companies using private capital with the goal of advancing maritime trade.

Portugal saw political and social upheaval between 1777 and 1822, first from the effects of the American Revolution and then from 1789 onwards from the influence of the French Revolution.

Aristocrats, merchants, and landlords primarily supported the English cause, while noblemen, progressive noblemen, intellectuals, and magistrates primarily adopted the French cause.

Portugal faced a classic strategic dilemma in these circumstances: either face a French invasion or lose its colonies to England. As a result of making Portugal's position clearer, the alliance with England was strengthened, which meant that the country had to put all of its resources, skills, and abilities into the military effort. This was the case in the Rossillian War, from 1793 to 1795, and in the French Invasions, from 1807 a 1811.

This conflict, part of Napoleon's plan to impose the Continental Blockade on the whole of Europe and thus defeat England, led to the transfer of the Portuguese court to Brazil in 1808, with the aim of ensuring dynastic continuity and thus guarantee Portugal's independence. When King João VI returned to Portugal in 1821, he found a different country, the result of the devastating marks left by successive years of war, as well as the effect of the transformation of the national political landscape, the result of the emergence of the nationalist and liberal movements that, in 1822, led to the independence of Brazil, the economic and maritime economic base of our empire. In this context, between the 17th and 18th centuries, Portuguese leaders, whether more or less charismatic and enlightened, either didn't have a maritime vision, or didn't have the capital opportunity to develop it, opportunity to develop and put it into practice: peace. From 1822 to 1974, the country tried to capture the resources in Africa that it had lost with independence from Brazil. However, the option was slow to come to fruition.

Between 1828 and 1834, there was a return to a liberal constitutional version of the Portuguese monarchy. Second, at the end of the 19th century, there was growing political unrest due to the increased expression and acceptance of Republican ideals. This unrest eventually led to the regicide of King Carlos in 1908 and the revolution of 5 October 1910, which established the Republic in Portugal.

Finally, the countries at the forefront of the industrial revolution were fighting over Africa and the sea, to the same extent and often in conflict with Portugal's interests.

Europe's geopolitical and geostrategic situation in the late 19th and early 20th century was very delicate, characterised by territorial disputes, economic rivalry and an incessant struggle for hegemony.

It was during this time that the major powers first tried to get along with each other, then they tried to form a coalition in order to protect their own interests and stop other countries or alliances from wanting to expand.

At this juncture, Portugal faced the political impasse that led to the formation of the Triple Alliance, a military coalition between Germany, Austria, and Italy based on the theories of Ratzel and Haushofer, and the Triple Entente, a military alliance between France, the United Kingdom, and Russia based on the theories of Mahan and Mackinder.

In this climate of deep ideological disputes, political guerrilla warfare, military coups, civil rebellions, and international conflicts, Portugal found itself involved in the First World War, trying at all costs to preserve its colonies.

This was followed, internally, by the military coup on 28 May 1926 that put an end to the First Republic and ushered in a period of military dictatorship, which ended with the approval of the Constitution of 1933. Professor Dr. António de Oliveira Salazar dominated the Estado Novo, an authoritarian single-party regime.

During his forty-year rule, António de Oliveira Salazar faced numerous challenges, including the Spanish Civil War from 1936 to 1939, World War II from 1939 to 1945, and the Overseas War from 1961 to 1968, which led to his replacement in government.On 25 April 1974, the 'Carnation Revolution' restored freedom and democracy to the Portuguese, and the country recognised the independence of its former colonies in Africa and Asia.

The successive conflicts that took place in the 19th and 20th centuries also did not give Portugal the opportunity or the conditions necessary to align and motivate cultural, political, economic, and security agents to return to the sea. However, from 1974 to the present day, the country has achieved political stability and contributed to internal tranquillization. In the context of security, as a member of NATO and the European Union, Portugal is enjoying lasting peace. On a cultural level, there has been an increase in the development of knowledge and expertise, which has contributed to achieving an unrivalled level of empowerment for Portuguese society.

Despite these circumstances, the country has experienced successive financial crises, which have posed enormous challenges to overcome.

In light of this, we must consider whether we should view these specific challenges as inevitable or as opportunities for national development.

From a maritime perspective, it seems that the recurring situation since 1974 should be seen as an opportunity to make Portugal grow economically. However, this naturally implies having the vision, courage, audacity, and determination to materialise a reunion of the Portuguese with the sea!

Without a doubt, the national relationship with the sea in the 21st century will significantly differ from that of the previous century.

In the past, the sea was, above all, a cultural cement of the national community, used as a means of travel, where users enjoyed a great deal of freedom of action.

It also served as a platform to project power for the interests of the most politically and economically dynamic countries.

Today, it is much more than that, as it serves as both a source of resources and the foundation of human life.

Despite the differences identified, it seems essential to continue to maintain some of the requirements of what was Portugal's first encounter with the sea, namely charismatic and enlightened leadership, capable of establishing a maritime vision, motivating citizens and citizens and aligning them with maritime projects, in everything that cultural, political, economic, security and environmental aspects. Considerations in this regard in the century stem from the industrial revolution, technological evolution and the use made of the sea in terms of industrial processes, which also pose industrial processes, which also pose new challenges for the country's with the sea.

The Reunion

In order for Portugal to reunite with the sea, it's important to identify what we need to do that we need to do in the times we live in, so that our descendants can enjoy it in the time when we won't be alive!

In this context, it is important to note that the sea in the 21st century presents the country with cultural, political, economic, environmental, and security challenges that must be met. To do this, the country's maritime policy must be made in a way that takes into account all of these factors We need this way of thinking not only to set and achieve nationalgoals but also to gather, organise, and use available resources, human and material abilities, and skills to carry out those goals.

On the other hand, it also implies a strategic approach that link and energise the various perspectives for dealing with maritime challenges, in the context of processes aimed at making the use of the sea viable the use of the sea in accordance with Portugal's development and security interests.

This dual approach, holistic and strategic, is essential in order to ensure balanced governance between the state, civil society, and the market at the local, regional, and international level. It is crucial to ensuring sustainability, meeting the needs of the present, and guaranteeing those of future generations. By its application, let's see how Portugal can overcome the challenges listed above and, in this way, reunite with the sea.

The cultural challenges have a philosophical meaning that has to do with intellectual life and thinking critically and reflectively about the sea. This can be seen in the study of the arts and sciences, which leads to a state of intellectual and moral perfection when it comes to the sea.

37

However, it also carries a sociological connotation, encompassing styles, methods, material values, and the moral virtues associated with the sea, as evidenced by the use of maritime objects and the adoption of maritime habits.

Overcoming the cultural challenges to Portugal's re-encounter with the sea sea implies deepening the mentality and strengthening the maritime will of the of the Portuguese, through the development of feelings, ideas and ways of and ways of feeling, as well as the intelligence to think about and understand the sea.

It also requires the development of resources, capacities, competences, and activities, as well as preserving cultural heritage, both tangible and intangible, and the maritime lifestyles of the Portuguese.

The political challenges stem from the fact that developed countries are seeking to develop or extend their borders at sea, while developing countries are seeking to preserve their interests in the face of the possibilities of expanding sovereignty that the United Nations Convention on the Law of the Sea has allowed for littoral states.

The developed countries are striving to assert their interests at sea, resulting in changes to the International Maritime Law. They view this action as a political act with economic consequences while simultaneously adopting the egalitarian rhetoric of the sea as a shared resource.

While the developing countries strive to assert their dominance over the sea, they perceive their diminished scientific, technological, and military capabilities as a threat to their interests, prompting them to seek compensation through external alliances.

Overcoming the political challenges to Portugal's reunion with the sea the sea fundamentally requires the mobilisation of civil society and the training of negotiators of maritime spaces. negotiators. It also requires the country to have the scientific and technological capacity to assert the national interest, combined with a multidisciplinary ability to occupy and use national maritime spaces for effective negotiation.

With regard to economic challenges, it is important to emphasise that, in the field of marine marine resources, they result from the exhaustion of living resources, the sea being the world's last great mine, the growing need for its resources and the rejection of measures based on eq On the other hand, in the field of commercial and industrial activities associated with the sea, the economic challenges are linked to the growing conflicts of interest between multiple national and international players, as well as the new opportunities associated with recreational boating, nautical tourism, maritime transport, ports and logistics, fishing, aquaculture, and the seafood fish industries, shipbuilding, and ship repair. We should strive for unity, solidarity, and sharing.

To overcome the economic problems that stand in the way of Portugal's return to the sea, a public policy for the economy of the sea must be made, scientists and engineers must be trained to find and use resources must be developed, and the authority at sea must be regulated and used in a coordinated way.

The environmental challenges primarily arise from issues related to sea-land interaction, pollution, and overexploitation of the oceans. These issues can lead to the degradation or destruction of marine ecosystems, which in turn affects humanity's survival. This is because life on the planet depends on the sea, which serves as an immense source of energy, water, and resources, supporting the survival of hundreds of millions of people and acting as the main stabiliser of the climate.

In this context, addressing the environmental obstacles to Portugal's reunification with the sea requires the regulation, monitoring, supervision, and control of maritime industrial activities.

containing the effects of environmental disturbances, especially maritime disturbances, especially maritime pollution; and using international mechanisms to limit mechanisms to limit the irrational use of the oceans

It is evident that international forums are already receiving criticism; there is diplomatic pressure and manipulation of public opinion in this regard.

It is inevitable that in the future, economic sanctions will be imposed for environmentally inconvenient uses, and eventually, military force will be used to prevent the misuse of the sea.

As for the security challenges, they are the result of erosive threats, namely transnational crime and trafficking in drugs, people, and weapons.

They are also a consequence of systemic terrorism and the proliferation of weapons of mass destruction.

The security challenges arise, threats resulting from natural accidents and other maritime accidents, and military threats, the most likely consequence of the more likely, the use of naval assets to achieve political ends, whether through low-intensity or submarine operations.

Let's focus on erosive and systemic threats, due to their insidious nature and higher probability of occurrence in failed states, jurisdictional waters States, jurisdictional waters, coastal lanes, maritime traffic and port security.

In this context, overcoming challenges to Portugal's reunion with the sea, recommends collaboration in strengthening international co-operation instruments, namely the security platforms, such as are, for example, the International Ship and Port Facility Security Code (ISPS Code), the Convention for the Suppression of Unlawful Acts and the Convention for the Suppression of Unlawful Acts against the Safety of Navigation (SUA 88) and the system (LRIT - Long Range Identification and

Tracking) (LRIT - Long Range Identification and Tracking). It also involves strengthening the instruments of international cooperation through international law, judicial mechanisms, and the use of force at sea.

It's also necessary to change how the government handles security at sea. This can be done by clearly planning public security capabilities, making the most of the potential for dual use of these capabilities, and coordinating the actions of the different public departments to create synergies.

Conclusion

In the past, Portugal encountered the sea due to circumstances resulting from the incorporation of maritime assets. These circumstances gradually awakened and fostered the national mentality and the national maritime will, creating opportunities for the country to carry out maritime activities that are crucial to its development and security.

The emergence of charismatic and enlightened leaders made it possible to a maritime vision for the country's future, which, combined with the ability to motivate citizens, mobilised the Portuguese for maritime projects that embodied the most epic feats in our collective history.

In contrast, the distancing from the maritime vision that governed national national life between the 14th and 16th centuries led to the unfeasibility national development and security interests that Portugal's vast maritime that Portugal's vast maritime spaces provided and required. The country focused on war and, progressively, on wanting everything. It lost everything over the course of the 16th to 20th centuries, when it was devastated by the effects of the dynastic crisis caused by the Alcacer-Quibir disaster: the loss of independence to Spain, the restoration of independence, the strengthening of the secular alliance with England, the French invasions, the transfer of the court to Brazil, the return of the court to Portugal, the political and social unrest that led to the establishment of the Republic, the struggle for European hegemony, deep ideological disputes, political guerrillas, military coups, and international conflicts.

This was the time when Portugal stopped thinking about the sea and put all of its resources, skills, and abilities into military projects to protect national interests or just to meet the obligations that came

42

from the strategic alliance that the country had to keep with England to protect its colonies and colonies and keep their independence.

The 21st century has finally brought peace, a fundamental opportunity for Portugal to explore the determinants and positive circumstances that will allow a new encounter with the sea to materialise!

A country needs to develop a way of life based on an integrated maritime vision that puts the sea at the centre of politics and inspires people to take on cultural, political, economic, and environmental challenges. This is something that happened in the 1400s and needs to happen again today.

To achieve this, it is essential to implement fifteen measures:

1. Develop feelings, ideas, and intelligence to think about and understand the sea.
2. Encouraging resources, capacités, compétences and maritime activities ;
3. Preserving tangible and intangible cultural heritage and maritime lifestyles;
4. Mobilising civil society on maritime issues and empowering negotiators of maritime spaces;
5. To have the scientific and technological capacity to assert the national interest;
6. Obtain multidisciplinary capacities to occupy and use national maritime national maritime spaces;
7. Drawing up a public policy on the economy of the sea;

8. Obtain scientific and technological capacity to identify and exploit inert resources;

9. To regulate and exercise, with coordination, the authority of the State at sea;

10. Regulate, monitor, supervise and control maritime activities;

11. Containing the effects of environmental disturbances, especially maritime pollution;

44

1. Use international mechanisms to limit the irrational use of the oceans;
2. Collaborate in strengthening maritime security co-operation;
3. Articulate the planning of public security capabilities, exploiting the potential of dual use:
4. Co-ordinate the actions of public departments, promoting synergies.

In essence, the fifteen measures listed help us realise the importance and scale of the challenges that Portugal will have to face in order to use the sea in the right way for its development and security interests.

However, in order for these measures to be recognised and accepted by the accepted by the Portuguese, awakening vocations and mobilising will, it is essential to continue to fulfil some of the requirements of what Portugal's first encounter with the sea, namely charismatic and enlightened leadership charismatic and enlightened leadership, capable of establishing a national maritime vision, to motivate the Portuguese and align them with maritime projects, cultural, political, economic, environmental and security aspects, environmental and security aspects, tasks in which we all need to get involved.

Without this leadership and without the collective commitment of the Portuguese to the sea, nothing significant will happen, despite the efforts made by remarkable people who, individually, have done a lot for Portugal!

António Silva Ribeiro - Admiral

The Sea: A Multi-faceted Approach

Artur Victoria

45

Forewords

The breathtaking view of the coastline is truly a sight to behold. As far as the eye can see, the endless stretch of land meets the ocean in perfect harmony. One can't help but admire how the traditional fishing villages seamlessly blend with the modern maritime infrastructure that dots the shoreline.

The picturesque landscape is a testament to the coexistence of old and new, a beautiful union of the past and the future. The scene is bustling with activity as fishermen go about their daily routine of tending to their nets. Robotic drones are tirelessly working alongside the fishermen to support sustainable aquaculture practices.

This perfect blend of traditional and cutting-edge technology is a true reflection of the community's adaptability and innovation. As the fishermen and drones work in unison, it's evident that they are not just coexisting but thriving together. Further out on the horizon, one can spot the towering structures of offshore wind farms and wave energy converters. These symbols of renewable energy are a testament to the community's commitment to harnessing the power of the sea in an environmentally sustainable way.

The sight of these structures proudly standing against the backdrop of the deep blue ocean is a true representation of the community's embrace of a greener and cleaner future.

The bustling port is in the middle of the image. The city is a melting pot of cultures, home to people from all over the world, making it a diverse and vibrant place. As you walk along the bustling streets of the city, you can see various marine technology start-ups that are revolutionising the maritime industry. The city is also home to some of the most prestigious research institutions, which are constantly pushing the boundaries of marine technology.

The harbour is a hub of activity, with a diverse array of vessels docking at the ports. From classic wooden boats to sleek research vessels, the

46

harbour is a reflection of the city's commitment to innovation and progress.

The city also boasts of its rich maritime heritage, which is evident in the various lighthouses and shipyards that stand tall alongside modern buildings. These landmarks serve as a reminder of the city's deep connection to the sea and its importance in shaping the city's identity. The blend of old and new in the city's architecture and industries creates a unique and dynamic atmosphere that is unlike any other city in the world.

On the shore, tourists flock to maritime museums and cultural centres, where they learn about the country's rich seafaring history and its important role in global exploration. These institutions offer a fascinating glimpse into the past, showcasing ancient artefacts and interactive exhibits that bring the country's maritime heritage to life. From ancient trading routes to modern-day naval forces, visitors gain a deep understanding of how the country has been shaped by its connection to the sea. However, while tourists are busy exploring the country's maritime past, marine conservation efforts are quietly

underway in the distance.

Scientists and activists are working tirelessly to protect fragile marine ecosystems for future generations. Whether it's through research, education, or advocacy, these individuals are dedicated to preserving the country's natural resources and promoting sustainable development. This commitment to conservation is a testament to the country's determination to balance economic growth with environmental responsibility.

As viewers take in this image, they can appreciate the country's deep-rooted connection with the sea. But they can also see the country's renewed efforts to leverage its maritime resources for sustainable development and economic growth.

This expansion offers a more complete picture of the country, highlighting both its rich history and its modern day initiatives. It also shows the seamless movement between the two, as the country continues to evolve while staying true to its roots. Ultimately, this image is a representation of the country's past, present, and future all at once.

To create a renewed image of the sea and its potential as an indispensable factor in identifying the economy, it's crucial to emphasize its significance in relation to Brazil's geographical, European, African, and Atlantic positions, as well as the vastness of the national maritime space. This vision should look toward the future with ambition and determination, placing the sea at the forefront of national ideology as a unifying force for the country.

The new image of the sea should be associated with modernity, innovation, mobilisation, wealth creation, power, and prestige. It should showcase Brazil and Portugal's commitment to leveraging their maritime resources for sustainable development and economic growth while honouring their maritime heritage.

This renewed image can be realized through concrete actions and initiatives that demonstrate the Portuguese-speaking countries' dedication to harnessing the full potential of the sea. Investments in maritime research, technology, and infrastructure can highlight the country's innovative approach to ocean exploration and exploitation. Collaborations with international partners can further enhance Brazil's position as a leader in maritime affairs.

Additionally, public awareness campaigns and educational programmes can help instil a sense of pride and appreciation for the sea among the Portuguese-speaking population, fostering cohesion around this shared maritime identity.

Overall, by replacing the sea at the heart of national ideology and promoting a vision of the future with broad horizons, Brazil and Portugal can position themselves as maritime powerhouses with the capabilities to thrive in the modern world.

Investing in education and increasing awareness about the strategic, economic, and social importance of the sea is crucial for fostering a more informed and intelligent relationship with it. This is especially true in today's world, where there is a growing need for sustainable practices and responsible use of our oceans.

By educating people about the economic opportunities presented by maritime industries such as shipping, fishing, tourism, and renewable energy, we can create a more sustainable and profitable future for both our economies and the environment. One important aspect of this education is to understand the potential of renewable energy in our oceans. With the growing demand for clean energy, the sea presents a vast potential for harnessing wind, wave, and tidal energy. By increasing awareness about these opportunities, we can encourage investment in renewable energy projects and reduce our dependence on fossil fuels. This will not only benefit the environment but also provide new job opportunities and boost economic growth.

Moreover, education about the sea can also promote responsible and sustainable fishing practises. By educating fishermen about the importance of maintaining fish stocks and avoiding overfishing, we can ensure the long-term viability of our oceans and the livelihoods of those who depend on them. Overall, investing in education and raising awareness about the sea can lead to a more sustainable and mutually beneficial relationship between humans and the ocean.

The sea is a vast and mysterious body of water, teeming with life and resources that are essential for our survival. To fully harness its potential and ensure its sustainable utilisation, society must have a deep understanding of the sea.

This understanding can be achieved through various efforts, such as continuous education, research, and communication. Education plays a vital role in cultivating this understanding. By educating scientists and academics, we can gain new insights and knowledge about the sea's complexities and challenges.

Politicians and investors also need to be educated about the sea, as they play a crucial role in making decisions that can have a significant impact on its health and resources. Students should also be included in this educational process, as they are the future leaders and decision-makers of society.

Furthermore, communication is key to building a comprehensive understanding of the sea. This includes reaching out to seafarers, journalists, and water sports enthusiasts who have firsthand experiences with the sea and can provide valuable insights. It is also important to include coastal communities and inland populations, as their livelihoods and well-being are directly affected by the state of the sea. By continuously educating and communicating with various sectors of society, we can achieve a better understanding of the sea and ensure its sustainable utilisation for generations to come.

Scientists and academics play a crucial role in advancing our knowledge and understanding of the marine environment. By fostering interdisciplinary research collaborations, we can gain invaluable insights into the intricate interactions between the ocean and human activities. These collaborations allow us to delve deeper into the complexities of marine science, technology, and policy, ultimately leading to more comprehensive and holistic perspectives. The ocean is a vast and dynamic ecosystem, and its study requires a multifaceted approach.

By bringing together experts from various fields, such as biology, chemistry, engineering, and sociology, a better understanding of the diverse and interconnected processes that shape the marine environment can be gained. This collaborative effort not only enhances our understanding of the ocean, but it also allows us to identify potential solutions to the complex issues facing our oceans, such as climate change, overfishing, and pollution. Interdisciplinary research collaborations also have the potential to bridge the gap between science and policy.

By involving policymakers in these collaborations, we can ensure that research findings are translated into effective and evidence-based policies to protect and sustain the oceans. This integration of science and policy is crucial for addressing the pressing environmental challenges facing our oceans and promoting sustainable use of marine resources. Therefore, fostering interdisciplinary collaborations is not only important for advancing scientific knowledge but also for creating positive impacts on the planet.

Marine conservation and sustainable development initiatives are critical for protecting our oceans and coastal communities. Policymakers and politicians must consider the science and engage with the stakeholders to develop working strategies of governance. Without proper management, marine resources will continue to decline, and coastal communities will suffer.

It is the responsibility of leaders to prioritise these issues and take action. Prioritising marine conservation and sustainable development initiatives involves implementing policies that promote the protection of marine resources and the communities that rely on them. This includes regulations to prevent overfishing, pollution, and other harmful practises.

In addition, policymakers must take into account the needs and opinions of stakeholders, such as fishermen and local communities, to ensure that their voices are heard and their livelihoods are protected.

Effective governance of marine resources and protection of coastal communities requires a multifaceted approach that takes into consideration both the environmental and social impacts. This means not only protecting marine ecosystems but also supporting the communities that depend on them. Prioritising marine conservation and sustainable development can ensure that oceans and coastal communities thrive for generations to come. It is time for leaders to take action and make this a top priority.

Investing in sustainable and eco-friendly maritime industries will help spur growth while keeping the planet safe. Increased investments in renewable energy and aquaculture will create jobs and promote eco-tourism, boosting local economies.

By supporting such projects, investors play a crucial role in protecting the oceans and preserving marine life. This will lead to a healthier and more sustainable planet for future generations.

Investors have the power to drive change and shape the future of maritime industries. Supporting innovative and sustainable projects can encourage the adoption of environmentally friendly practises in these industries. This will not only benefit the environment but also create a competitive edge for these businesses in the long run.

Furthermore, investing in sustainable maritime industries can have a positive impact on local communities by providing them with alternative sources of income and encouraging social and economic development. In addition to promoting economic growth and preserving the environment, investing in sustainable maritime industries can also have a positive impact on global trade. With increasing concerns about climate change and environmental degradation, there is a growing demand for sustainable products and services.

By supporting eco-friendly maritime businesses, investors can tap into this growing market and contribute to a more sustainable global trade system. As a result, investing in sustainable maritime industries not only benefits investors and local communities but also has a larger impact on the global economy and the planet as a whole. Students should be highly encouraged to pursue education and careers in marine science, as it is an area of study that is not only fascinating but also essential for the future of the planet. In order to truly understand and protect our oceans, we need a new generation of thinkers and problem-solvers who are passionate about marine life and its preservation.

By providing students with the necessary tools and resources to succeed in this field, we are not only shaping the future of our oceans but also preparing the next generation of leaders and innovators in the maritime sector.

Furthermore, marine engineering careers are becoming increasingly important as our reliance on ocean-based resources grows. From offshore wind farms to sustainable aquaculture, there is a growing demand for engineers who can design and implement solutions to the challenges facing our oceans.

Encouraging students to pursue this field of study not only offers them exciting and fulfilling career opportunities, but also ensures responsible and sustainable management and utilization of our oceans.

In addition to marine science and engineering, it is also crucial to encourage students to pursue careers in marine management. As the oceans face ever-increasing threats from human activities, it is essential to have skilled and knowledgeable individuals who can effectively manage and protect these valuable resources. By promoting education and career opportunities in marine management, we are ultimately investing in the future health and sustainability of our oceans for generations to come.

Seafarers will see the benefits of training programmes and awareness campaigns that focus on safety and responsible recreation. These efforts will help ensure a greater understanding of the environment and how to best protect it.

Training programmes can help seafarers prepare for the water and provide them with the knowledge they need to stay safe. Additionally, these campaigns will also help promote environmental stewardship and encourage responsible water sports activities. This will lead to a more sustainable use of water resources and help maintain the natural beauty of water bodies.

With proper training and awareness, water sports enthusiasts can learn how to minimise their impact on the environment and become responsible water stewards. Furthermore, these programmes and campaigns will also improve the overall safety of water sports and recreational activities.

By educating seafarers and water sports enthusiasts on the potential risks and how to mitigate them, accidents and injuries can be significantly reduced. This will benefit not only the individuals participating in these activities but also the overall safety of the waterways.

In conclusion, the implementation of training programmes and awareness campaigns will have a positive impact on both individuals and the environment, leading to a more enjoyable and sustainable experience on the water.

Journalists are often our lifeline to the world, bringing us information and news from far-off places that we may never get to see for ourselves. They play an important role in our society, especially when it comes to raising awareness about issues that impact our planet. One such issue is the state of our oceans and marine life, which are constantly threatened by human activity.

By shining a light on the challenges facing our seas, journalists have the power to educate the public and drive change. From overfishing to plastic pollution, there are numerous threats to marine life that need to be addressed.

Through their reporting, journalists can bring these issues to the forefront and hold those responsible accountable. They also have the ability to showcase the opportunities that exist for preserving our oceans, such as sustainable fishing practises and marine conservation efforts. In addition to informing the public, journalists also have the power to inspire action and bring about positive change.

They can motivate individuals and governments to take steps to protect and preserve our oceans for future generations by highlighting their importance and fragility. We, as a society, must recognise the critical role that journalists play in keeping us informed about marine issues and supporting their efforts to raise awareness and drive change.

Coastal communities are uniquely poised to lead conservation efforts in their region. They can take a more hands-on approach to preserving their local environment by engaging in community-based initiatives. Activities like coastal cleanups and educational outreach programs can accomplish this.

By fostering a sense of connection and stewardship toward the marine environment, inland populations can also play a crucial role in conservation efforts. Educational outreach programmes are especially important, as they can help educate the public on the importance of preserving the marine environment and the impact their actions can have. Through these people can learn about the various threats facing the oceans and what they can do to make a positive difference. This can include simple actions such as reducing their use of single-use plastics or participating in local coastal cleanups.

Furthermore, by involving both coastal and inland populations in these initiatives, a more complete picture of the interconnectedness of the marine environment can be understood. This can help bridge the gap between these two communities and create a unified effort toward conservation. By staying on track with these efforts and knowing when to quit, we can ensure that our oceans are protected for generations to come.

By encouraging individuals from all backgrounds to engage in the ongoing education and exploration of the ocean, we can foster a greater understanding and reverence for its intricate ecosystems.

This shared knowledge can lead to collaborative efforts to preserve and safeguard the ocean for the benefit of future generations. Through inclusive learning opportunities, we can create a sense of ownership and responsibility for the ocean, motivating individuals to take action for its preservation. Furthermore, by promoting continuous learning and understanding of the sea, we can cultivate a deeper appreciation for its value and significance in our lives.

The ocean not only provides us with vital resources and supports countless industries, but it also plays a crucial role in our planet's health and well-being. By expanding our understanding of the ocean's importance, we can better appreciate the need for sustainable management and conservation practices. This, in turn, can lead to more effective and collaborative efforts to protect and preserve this vital resource for the benefit of current and future generations.

The sea is not only a geographical feature but also a source of national pride. It symbolises the long-standing relationship of the country with the sea, and its vastness and depth inspire a sense of awe and respect. Its presence is felt in every aspect of life, from the economy to cultural traditions.

The sea holds a special place in the hearts of the people, serving as a reminder of their seafaring ancestors and the country's rich maritime history.

The sea is a powerful image that represents the country's connection to the rest of the world. Its vast expanse serves as a bridge between the nation and other lands, allowing for trade and cultural exchange.

It also serves as a source of livelihood for many, with fishing being a major industry. The sea's influence can be seen in the country's cuisine, music, and art, all of which have elements of the sea woven into them. The sea is not just a physical feature; it is also a cultural and emotional one, evoking a sense of nostalgia and pride in the people.

The sea offers a vast expanse of opportunity for Brazil, Portugal, and the Portuguese-speaking countries. It connects them to the rest of the world, allowing for trade and commerce to flourish. The sea is a gateway to international collaboration and exchange, providing a platform for these countries to showcase their talents and products.

This has opened up new avenues for growth and development, boosting the economies of these nations. Portugal's maritime history is a testament to its seafaring prowess. The Portuguese were pioneers of exploration and discovery, venturing into uncharted waters and expanding their horizons.

This legacy has played a crucial role in shaping Portugal's reputation as a maritime nation. The country's expertise in navigating the seas and rich seafaring culture has made it a leader in the global maritime industry. It continues to be a major player in the international shipping market, providing a vital link between continents and facilitating global trade. The sea has truly transcended its physical boundaries, offering endless possibilities and opportunities to Brazil, Portugal, and the Portuguese-speaking countries. It has connected these nations to the rest of the world through its vast and powerful waters, allowing them to thrive and prosper. Portugal's maritime history and expertise have only added to the sea's grandeur, cementing its status as a vital lifeline for these countries. The sea continues to be a source of inspiration and growth, driving these nations toward a brighter future.

A deep dive into the sea can unveil hidden treasures of ambition, aspiration, and forward-thinking mentality. It serves as a symbol of endless opportunities waiting to be unlocked for economic growth, innovation, and national development.

The vast expanse of the sea beckons us to think big, dream bigger, and embrace its potential with open arms. As we set sail on the sea of aspirations, we must remember to navigate it sustainably. Just as we must respect the sea and its resources, we must also ensure that our actions align with the greater good.

We can benefit not only ourselves but also future generations by responsibly leveraging the ocean's resources. The sea is not just a symbol of ambition but also a reminder of our responsibility toward the environment and our fellow beings. The sea's potential is limitless, and our ambition to harness it must be matched by our determination to protect it.

As we venture into the depths of the sea, let us not forget the purpose behind our journey. Let us strive to create a more complete and sustainable picture where economic growth goes hand in hand with environmental conservation. The sea calls out to us, and it is up to us to answer its call with purpose and responsibility.

Outdated attitudes and beliefs hinder the realization of the sea's vast potential.

To truly harness the power of the sea, a shift in mindset is necessary.

We must abandon our narrow and restrictive views of the ocean and embrace a culture of maritime literacy. This includes educating ourselves and others on the strategic and economic importance of the sea, as well as investing in industries and infrastructure that will allow us to fully utilise its resources. The sea not only holds immense economic value but also serves as a vital strategic asset.

As countries around the world compete for resources, the sea has become a key battleground. It is crucial that we recognise and prioritise the significance of the sea in global affairs. By promoting awareness of its strategic importance, we can work toward developing policies and initiatives that will safeguard our interests and maintain our position in the global economy. This requires a concerted effort from all stakeholders, including governments, businesses, and individuals, to foster a more comprehensive understanding of the sea and its potential. To fully capitalize on the sea's potential, we must also invest in maritime industries and infrastructure.

This includes developing new technologies and methods for sustainable resource extraction, as well as improving transportation and communication networks. By doing so, we can create new opportunities for economic growth and increase our ability to navigate and utilise the vast oceanic expanse. It is time for a paradigm shift in our thinking about the sea, and by promoting a culture of maritime literacy and investing in its industries and infrastructure, we can unlock its full potential and reap the rewards for generations to come.

By harnessing the full potential of the sea, the Portuguese-speaking countries can greatly increase their global presence and prosperity. The vast expanse of the ocean provides ample opportunity for growth in trade, travel, and even scientific research.

By embracing this fundamental element of their identity, the Portuguese-speaking countries are able to tap into a wealth of possibilities. With a newfound focus on the sea, these countries can truly flourish on the global stage and establish themselves as maritime powerhouses. The incorporation of the sea into their identity also allows for a more unified sense of purpose among the Portuguese-speaking countries.

By sharing this common thread, they are able to come together in pursuit of a shared goal. This sense of unity and shared vision is crucial in navigating the complexities of the global stage.

These countries are able to create a strong and lasting bond by embracing the sea as a central element of their identity, allowing them to thrive in the face of challenges and competition. Furthermore, by integrating the sea into their plans for the future, the Portuguese-speaking countries are able to diversify their economies and reduce their reliance on a single industry. This opens up new opportunities for growth and development, allowing for a more sustainable and resilient future.

By staying true to their maritime roots, these countries are able to evolve and adapt, ensuring their continued success on the global stage. Ultimately, by embracing the sea as a central element of their identity and vision for the future, the Portuguese-speaking countries are able to reach new heights and truly make their mark in the world. The sea has always been a source of both mystery and wonder.

It has been a place where societies have flourished and cultures have been born. However, in recent years, the health and vitality of the sea have been threatened by pollution and overfishing. As a result, it is now more important than ever to create a new vision for the sea, one that takes into account both cultural and societal change.

This vision must be holistic and should encompass a range of strategies that can be implemented across multiple sectors of society. This process necessitates a renewed appreciation for the sea's potential. By recognizing the value of the sea, both in terms of its resources and its cultural significance, we can begin to foster a deeper understanding of the sea's role in our lives. This appreciation must go beyond economic considerations and include the social and environmental benefits that the sea provides.

Additionally, a commitment to fostering innovation and proactive engagement across civil and political spheres will be key to creating a new vision for the sea. By involving diverse stakeholders and promoting collaboration, we can ensure that this vision is inclusive and sustainable. In conclusion, creating a new vision for the sea is a complex and multifaceted endeavour that requires a comprehensive approach. By recognising the need for cultural and societal change, appreciating the potential of the sea, and encouraging innovation and proactive engagement, we can begin to create a more sustainable and prosperous future for the ocean.

The collaboration between industry, academia, and government is crucial in creating opportunities for innovation and entrepreneurship in the maritime sector. As the sea presents itself as a vast and largely untapped resource, entrepreneurs and investors play a vital role in driving this transformation forward.

By taking the lead in spearheading initiatives that capitalise on these opportunities, they can break the existing vicious circle where a lack of investment leads to limited opportunities. To truly address this challenge, it is essential to foster an ecosystem that not only supports but also incentivises innovation and entrepreneurship in maritime industries. This may include providing financial incentives to encourage investment as well as access to resources and infrastructure to facilitate the development of new ideas and initiatives. We can create a more conducive environment for young people to pursue careers in these industries, leading to a more diverse and vibrant workforce.

Furthermore, this coordinated effort can also help address the issue of sustainability in the maritime industry. With a focus on innovation and collaboration, we can find new ways to reduce the environmental impact of these sectors and promote sustainable practises. This not only benefits the health of our oceans but also presents new opportunities for entrepreneurs and investors to capitalise on. We can drive the transformation necessary to create a more sustainable and prosperous maritime industry by fostering an ecosystem that supports and incentivizes innovation and entrepreneurship.

Enlisting the support and participation of young people in the maritime sector can also play an important role in promoting sustainable development and conservation efforts. By involving youth in ocean conservation initiatives, we can tap into their passion and innovative thinking to find solutions to pressing environmental challenges. This can also help foster a greater sense of connection and understanding between younger generations and the ocean, ultimately leading to more responsible and sustainable use of marine resources.

In addition, it is crucial for maritime industries to embrace diversity and inclusivity in their workforce. This means actively recruiting and promoting individuals from diverse backgrounds, including women and minority groups. By creating a more inclusive workplace, companies can benefit from a wider range of perspectives, ideas, and talents, ultimately leading to increased innovation and success.

It is also important for companies to provide equal opportunities for career advancement and support for work-life balance for all employees, regardless of gender or background. Lastly, promoting maritime careers and fostering a culture of marine literacy can also have positive economic impacts for Portuguese-speaking countries.

By investing in education and training programmes, as well as promoting the diverse career opportunities in the maritime sector, we can cultivate a skilled and knowledgeable workforce that can drive innovation and growth in the industry. This can also attract international talent and investment, positioning Portugal as a leader in the maritime world. Therefore, efforts to promote and support maritime careers should be seen as a crucial component of Portugal's economic development strategy.

1 - Water sports, pleasure boating, and nautical tourism

Water sports, pleasure boating, and nautical tourism have been known to have a great impact on a country's economy and strategic options.

These activities, which have the potential to attract tourists and provide a diverse range of services, have been instrumental in increasing the country's appeal as a tourist destination. These industries have not only created job opportunities for local residents, but they have also helped to promote the country's cultural heritage and natural beauty to the rest of the world.

Water sports, pleasure boating, and nautical tourism all have an impact on the development of infrastructure and facilities to support these activities. Marina construction, beach resorts, and other recreational facilities have been built to cater to the growing demand for these activities.

This has not only improved the country's tourism sector, but it has also boosted its overall development. Furthermore, promoting eco-friendly practices in these industries has contributed to the conservation of the country's natural resources, making it a more sustainable and attractive destination for tourists.

The potential for growth and expansion in these industries is immense, and it may lead to even greater opportunities for the country in the future. By continuously improving and diversifying the range of tourist services available, the country can attract a wider range of tourists and maintain its position as a top tourist destination.

With the right balance of economic and environmental sustainability, water sports, pleasure boating, and nautical tourism

63

can continue to positively impact the country's strategic options and contribute to its overall development.

These activities are not only beneficial for the tourism sector but also for the local economy. With the wide variety of activities available, tourists can partake in everything from extreme water sports to peaceful, relaxing experiences on the water. By appealing to a diverse range of travellers, the tourism industry sustains its strength throughout the year rather than relying solely on peak seasons. This, in turn, helps increase overall tourism revenue, which has a significant impact on the local economy. Furthermore, these activities also have a positive impact on the environment. With an increase in eco-tourism, tourists are able to experience the beauty of nature while also learning about the importance of conservation and sustainable practises. This not only helps to preserve the natural resources for future generations but also supports the local community by promoting responsible tourism. By offering a diverse range of activities, the tourism industry is able to attract more environmentally conscious and socially responsible travellers. This not only benefits the environment but also helps enhance the overall image and reputation of the tourism sector.

Enhancing the overall experience for tourists, the beautiful and diverse coastlines of the CPLP countries offer an array of water sports and nautical activities. With miles of pristine beaches and a friendly climate, visitors can partake in memorable experiences that showcase the natural beauty and rich maritime heritage of these countries. Whether it be surfing in the powerful waves of the ocean or leisurely sailing along the picturesque coastline, there is no shortage of opportunities for water-based recreation. Tourists can indulge in a variety of water sports and activities, each providing a unique perspective on the country's coastal charm. For the adventurous, there are thrilling options such as jet skiing, parasailing, and scuba diving, which allow them to explore the depths of the crystal-clear waters and encounter diverse marine life.

For those seeking a more relaxing experience, the calm and tranquil waters are perfect for activities like paddle boarding, kayaking, and snorkelling. No matter the preference, the coastline of the CPLP countries offers something for everyone to enjoy. Moreover, the water sports and nautical activities not only showcase the natural beauty of the CPLP countries but also provide insights into their rich maritime heritage.

Many of these activities have been passed down for generations, allowing tourists to immerse themselves in the local culture and traditions. From the skilled fishermen casting their nets to the expert surfers riding the waves, visitors can witness and appreciate the deep connection these countries have with the ocean. Overall, the extensive coastline and favourable climate in the CPLP countries make it a prime destination for tourists seeking unique and unforgettable water-based experiences.

The development of infrastructure and services catering to water sports and nautical tourism is not only beneficial for tourists but also for the local communities. It can stimulate economic growth by creating new job opportunities and supporting local businesses. With a growing number of tourists choosing coastal destinations for their vacations, the demand for various water activities has also increased.

Marinas, sailing schools, equipment rental shops, and waterfront restaurants are just a few examples of businesses that can thrive in these areas. The influx of tourists drawn to these activities not only boosts the local economy but also creates a ripple effect in other sectors. For instance, the need for supplies and maintenance services also increases, providing employment opportunities for the locals.

Furthermore, the development of water sports and nautical tourism promotes the preservation and conservation of natural resources. The ocean serves as the focal point for these activities, inspiring individuals to care for the marine environment.

This can also open up opportunities for environmental initiatives and projects that can benefit the community in the long run. Therefore, investing in the development of such facilities can have a positive impact on both the economy and the environment, making it a win-win situation for everyone involved.

Promoting water sports and nautical tourism not only attracts visitors to the country, but it also aligns with the broader strategic objectives of sustainable development and environmental conservation. By promoting responsible practises and eco-friendly initiatives within these sectors, we can showcase our commitment to preserving our marine ecosystems while still reaping the economic benefits of tourism. This can also serve as a model for other countries to follow, as responsible tourism becomes increasingly important in today's world.

Furthermore, the development of water sports and nautical tourism can have a positive impact on the local economy. It can create job opportunities for the local community, especially in coastal areas where traditional livelihoods may be declining. This can lead to a more sustainable and diverse economy, ultimately benefiting the country as a whole.

Furthermore, by promoting these activities, we can encourage the preservation and protection of the country's natural resources, which are critical for the tourism industry's sustainability. In order to fully achieve the potential of water sports and nautical tourism, it is important to also address any negative impacts that may arise. This may involve implementing regulations and guidelines to protect marine life and promote responsible behaviour among tourists.

By doing so, we can ensure the long-term sustainability of these activities and maintain the balance between economic benefits and environmental conservation. With a holistic approach, promoting water sports and nautical tourism can truly align with the country's broader strategic objectives of sustainable development and environmental conservation.

Marina infrastructure, facilities, and services are essential for countries to position themselves as preferred stopover destinations for sailing enthusiasts. By investing in such resources, countries can encourage longer stays, providing a major boost to their tourism revenue. The strategic location of these marinas along popular sailing routes makes them an ideal spot for travellers to stop their journeys, rest, and explore their destinations.

Moreover, the development of marinas can also open up opportunities for local businesses, creating jobs and promoting economic growth. The availability of modern facilities and services at these marinas can attract more visitors, leading to a significant increase in tourism revenue. In addition, marinas can also provide an array of leisure activities, such as water sports, fishing, and sightseeing, making them attractive destinations for tourists seeking unique experiences.

Investing in marina infrastructure is beneficial not only for the countries but also for the sailing community. With safe and well-equipped marinas available, sailors can confidently plan their journeys, knowing they have a reliable stopover destination. This not only adds convenience but also enhances the overall sailing experience. Overall, the development of marinas can have a positive impact on both the countries and the sailing community, making it a valuable investment for all parties involved.

One of the biggest draws for boat owners is the availability of permanent and seasonal parking options for recreational boats. Owners seeking to safeguard their investment can find peace of mind in these facilities, which provide a safe and convenient place to dock their boats.

In addition, these facilities often offer attractive packages for hibernation periods, making it easier and more cost-effective for boat owners to store their boats during the off-season. Along with storage options, these facilities also provide maintenance services for boats, ensuring that they are in top condition when owners are ready to take them out on the water. This can include routine maintenance, repairs, and upgrades, all done by experienced professionals.

And, as an added bonus, many of these facilities also offer access to amenities such as fuelling stations, restaurants, and shops, making it a one-stop destination for all of a boat owner's needs. By providing these services and amenities, countries can incentivise boat owners to choose them as their preferred base for exploring ocean waters. This not only boosts the local economy, but it also allows boat owners to fully enjoy their time on the water without having to worry about the logistics of storage and maintenance. Overall, offering these options caters to boat owners' needs and can greatly benefit both the country and the boating community.

The sea is an ever-changing landscape that has shaped the culture and lives of coastal communities for centuries. From the awe-inspiring power of the ocean's waves to the serene beauty of the shoreline, the sea offers a multitude of experiences that are perfect for travellers seeking a deeper connection with nature.

Eco-tours, for example, provide a unique opportunity to explore the diverse ecosystems and marine life that call the sea their home. These tours allow travellers to not only witness the beauty and wonders of the ocean but also learn about the importance of conservation and sustainability.

Coastal excursions, on the other hand, offer a more relaxed and immersive way to experience the sea. Whether it's taking a leisurely stroll on the beach, trying out water sports like surfing or paddle boarding, or simply enjoying a seafood feast by the shore, these activities allow travellers to fully embrace the coastal way of life. They also offer a chance to connect with the local community and learn about their traditions and customs, making for a more authentic and enriching travel experience. With the sea as a backdrop, these excursions provide the perfect balance between relaxation and cultural immersion for travellers seeking a dynamic and meaningful trip.

From the moment you step into this ancient town, you are immediately surrounded by a rich heritage of culture and gastronomy.

Along the coastline, there are numerous historic maritime landmarks waiting to be explored, each with its own unique story to tell. As you wander through the cobblestone streets, you will come across quaint fishing villages where you can witness the traditional way of life still being preserved by the locals.

This town's maritime culture is also reflected in its cuisine. Seafood is a staple in the local diet, and there is no shortage of delicious dishes to try. From freshly caught fish to succulent shellfish, every meal is a mouth-watering experience.

These lively celebrations are a chance to experience the local customs and traditions firsthand, and they are not to be missed. A trip to town offers more than just a vacation; it offers a journey through time and traditions. With its fascinating history, delicious cuisine, and vibrant festivals, this destination truly offers a holistic and enriching travel experience. So pack your bags and get ready to immerse yourself in this maritime paradise.

To promote tourism, the country must develop activities that revolve around renting equipment and recreational boats. By doing so, the countries will be able to attract tourists who are interested in experiencing the country's coastal waters through various water sports. This can be achieved by offering rental services for equipment such as surfboards, kayaks, paddleboards, and sailboats.

Moreover, providing training in recreational boating and water sports is also crucial to fostering a thriving nautical tourism industry. With proper training, tourists can safely and confidently participate in water activities, further enhancing their overall experience in the destination country.

This can also open up job opportunities for locals who can serve as instructors and guides, thereby contributing to the country's economy. Furthermore, developing activities related to equipment and recreational boat rental can help promote sustainable tourism. Tourists do not need to purchase their own equipment because rental services are provided, reducing waste and promoting environmental conservation.

This can also attract eco-conscious travellers who are looking for ways to minimise their carbon footprints while still enjoying water activities. With these efforts, the country can establish itself as a top destination for nautical tourism, attracting a diverse range of tourists and boosting its economy.

There are numerous benefits to providing such training and certification programmes. Not only do these help to create a safer and more responsible recreational boating and water sports culture, but they also have the potential to boost the local economy. By offering these courses, certified instructors and training centres are able to attract tourists with different levels of experience and thus generate revenue for the community.

In addition, these programmes also serve as an opportunity for locals to develop and refine their own skills and knowledge in recreational boating and water sports. This can lead to an increase in job opportunities as well as a sense of pride and ownership in the community's water-based activities. Furthermore, having a well-trained and certified population of tourists and locals can help reduce the number of accidents and incidents that occur on the water, thereby ensuring a safer and more enjoyable experience for all.

Overall, offering these training and certification programs benefits both tourists and the community at large.

Land-based support services are crucial for creating a well-rounded experience for those who visit the beach for sports and water-based activities. These amenities can greatly enhance the overall experience for visitors, making it more convenient and enjoyable.

Access to facilities such as changing rooms, showers, and rest areas allows for a seamless transition from land to water, making it easier for sportspeople to engage in their activities. Additionally, having access to restaurants, cafes, and shops selling beachwear and supplies provides a convenient and enjoyable way for nautical tourists to spend their time when they are not on the water. Moreover, these land-based support services also benefit the local economy by providing job opportunities and boosting tourism. These facilities and amenities can attract more visitors to the area, resulting in increased revenue for local businesses. This, in turn, can lead to the development and growth of the surrounding community. Additionally, these services can also help promote sustainable tourism by providing environmentally friendly options for visitors, such as eco-friendly restaurants and shops. In conclusion, land-based support services play a crucial role in enhancing the overall visitor experience and contributing to the local economy.

These services can attract more tourists and promote sustainable tourism by offering convenient and enjoyable amenities. They also play a vital role in creating a well-rounded experience for sportspeople and nautical tourists, making it easier for them to engage in their activities and enjoy their time at the beach.

Cruise ship tourism is a valuable asset that allows maritime cities to introduce their unique and diverse cultures to a wide range of visitors. By developing infrastructure and services to accommodate cruise ships, these cities can take advantage of their coastal assets and showcase their local attractions. The development of docking facilities, passenger terminals, and shore excursions will provide a seamless and memorable experience for all cruise passengers.

The benefits of cruise ship tourism are plentiful. It not only brings in revenue for the cities, but it also allows for cultural exchange and exposure. With the development of docking facilities, maritime cities can attract a diverse range of visitors from all over the world. This, in turn, can stimulate local economies and promote growth in the tourism industry. Cruise passengers can explore the coastal cities and attractions, immersing themselves in the unique cultures and experiences that each city has to offer.

Furthermore, the development of infrastructure and services for cruise ships can lead to job creation and economic growth in cities. As the demand for cruise ship tourism increases, so does the need for workers to support the industry. This can provide job opportunities for locals and stimulate economic growth in the region. Overall, cruise ship tourism presents a valuable opportunity for maritime cities to showcase their assets, attract visitors, and promote economic growth.

The first and foremost step in responding to current and potential domestic demand is to understand the diverse preferences and needs of domestic travellers.

This requires a multifaceted approach that takes into account the different segments of water sports, recreation, and tourism. Each segment has its own unique set of requirements, and addressing them effectively is crucial to meeting the demand. One of the key factors to consider is the changing trends in water sports, recreation, and tourism.

As the tastes and preferences of domestic travellers evolve, it becomes important to adapt and offer new, exciting experiences.

This can help attract a wider audience and ensure that the demand is met.

Moreover, understanding the target market and their expectations is equally important.

By catering to their specific needs, the industry can create a loyal customer base and foster growth. Another crucial aspect to consider is the role of technology in the water sports, recreation, and tourism industries. With the rise of social media and online platforms, it's become easier for domestic travelers to access information and plan their trips. Hence, integrating technology into the industry is essential for reaching out to potential customers and providing a seamless experience. By staying up-to-date with the latest technological trends, the industry can enhance the overall customer experience and meet the ever-changing demands of domestic travellers.

Strategies to consider:

Market research and segmentation are critical components of any successful business venture. Before launching a product or service, it is imperative to gain a deep understanding of the target market and their preferences. In the case of water sports, recreation, and tourism, conducting thorough market research is even more crucial due to the diverse and ever-evolving nature of this industry.

Through market research, businesses can gain insights into the preferences, behaviours, and motivations of their target audience. This information can then be used to segment the market based on various factors, such as age, income, interests, and geographic location. By dividing the market into smaller, more homogeneous groups, businesses can create tailored offerings that cater to the specific needs of each segment.

For instance, a company offering water sports activities in a coastal town may choose to target younger audiences with a higher income who are interested in adventure and thrill-seeking activities. On the other hand, a company situated in a landlocked state may focus on families with children who are looking for more relaxed and family-friendly water activities. By understanding the unique needs and preferences of each segment, businesses can create customised marketing strategies and offerings that resonate with their target audience, ultimately leading to increased sales and customer satisfaction.

As the tourism industry continues to grow, so too must the range of products and experiences available to travellers. This expansion not only caters to a diverse range of customers but also serves to develop and boost domestic demand. By offering package deals for families, adventure tours for thrill-seekers, wellness retreats for relaxation, and cultural experiences for history enthusiasts, businesses can attract a wide variety of customers and establish themselves as leaders in the industry.

Creating a range of products and experiences also allows for and personalisation, which enhances the overall customer experience. Families can choose from a variety of package deals that cater to their specific needs and preferences, while thrill-seekers can opt for adventure tours that push their limits. Wellness retreats provide a much-needed escape for those seeking relaxation, while cultural experiences allow history enthusiasts to immerse themselves in the local culture. By catering to different segments of domestic demand, businesses can stay ahead of their competition and continuously attract new customers.

Moreover, developing a diverse range of products and experiences also contributes to the overall growth and sustainability of the tourism industry. By offering a variety of options, businesses can attract various types of travellers and encourage them to explore different destinations. This not only promotes economic growth but also helps preserve and protect the natural and cultural resources of each destination. By following these important instructions and expanding in a purposeful manner, businesses can thrive in the ever-evolving and competitive field of tourism.

Infrastructure can be a powerful driving force for growth and development in a community. Investing in infrastructure, especially for water sports, recreation, and tourism activities, can have a significant impact on the local economy.

By developing and enhancing infrastructure, such as marinas, promenades, beach facilities, and recreational areas, a community can attract more tourists, boost local businesses, and create new job opportunities. Moreover, investing in infrastructure can also improve residents' overall quality of life. With better facilities and recreational areas, there will be more opportunities for people to engage in water sports and other recreational activities, leading to a healthier and more active community. This can also contribute to the growth of local businesses, as more people will be drawn to the area for its recreational opportunities. In addition, infrastructure development can also have environmental benefits. With proper planning and design, these developments can help preserve and protect natural resources, such as beaches and water bodies.

By providing designated areas for water sports and recreation, it can prevent overcrowding and minimise the impact on the environment. In turn, this can attract more tourists who are looking for sustainable and eco-friendly destinations. Overall, investing in infrastructure development can have numerous positive effects on a community, making it a worthwhile investment.

Training and education programmes are essential for local businesses and service providers to create quality experiences for domestic travellers. By providing training in customer service, safety standards, environmental stewardship, and cultural sensitivity, businesses can ensure that their employees are equipped with the necessary skills to offer the best possible services to their guests. This not only leads to satisfied customers but also helps in building a positive reputation for the business.

: Furthermore, these training and education programs significantly contribute to the promotion of sustainable and responsible tourism practices.

By educating businesses and service providers about environmental stewardship, they can learn to minimise their impact on the environment and contribute to preserving local cultures and traditions. This, in turn, contributes to a more authentic and immersive travel experience that many tourists highly value.In addition to this, training and education programmes also benefit the local community as a whole.

By providing these opportunities, businesses and service providers can create employment and economic growth in the area. This not only helps improve the standard of living for the locals but also creates a positive relationship between the community and the tourism industry. Therefore, it is crucial for destinations to invest in training and education programmes to ensure the sustainability and success of their tourism industry.

Promotion and marketing campaigns are a vital aspect of reaching potential travellers and raising awareness of available water sports, recreation, and tourism offerings. Utilising various channels, such as social media and digital advertising, can be extremely effective in targeting domestic audiences. By reaching out to these audiences, they will be more aware of the available recreational activities and more likely to visit the area. We can utilize tourism websites and local events, in addition to social media and digital advertising, to promote and market water sports, recreation, and tourism offerings.

These channels can provide a more comprehensive and detailed view of the available activities, which makes it easier for potential travellers to plan their trips. Participating in local events, such as festivals or sports tournaments, can help the area showcase its offerings and attract a larger audience. Implementing targeted promotion and marketing campaigns not only raises awareness among domestic audiences, but it also helps boost the local economy. By attracting more travellers, the area will see an increase in tourism revenues, which can be beneficial for the community as a whole. By utilizing various channels and staying on track with the campaign's purpose, the area can effectively promote its water sports, recreation, and tourism offerings, resulting in a significant increase in visitors. Knowing when to quit and ending the campaign once its goals have been achieved is essential to avoiding overspending and

maintaining the success of the promotional efforts.

In order to foster collaboration and partnerships between public and private sector stakeholders, it is essential to establish strong connections between various organisations and groups. This can include tourism boards, local businesses, recreational organisations, and community groups.

By encouraging teamwork and collaboration, these stakeholders can combine their resources and expertise to improve the overall visitor experience. Collaborative efforts like event organization, local attraction promotion, and unique visitor experiences can achieve this.

One of the key benefits of collaboration and partnerships is the ability to leverage resources and expertise. By bringing together different perspectives and strengths, stakeholders can enhance the quality of services and products offered to visitors. This can result in a more diverse and attractive range of options for tourists, ultimately contributing to the growth and sustainability of the tourism industry.

Additionally, partnerships can also help to minimise costs and risks, as organisations can share the burden of planning and implementing initiatives. It is important to note that collaboration and partnerships should not be limited to just the tourism industry but also include other sectors such as local businesses and community groups.

By involving a diverse range of stakeholders, it allows for a more well-rounded and holistic approach to promoting tourism. Furthermore, by building strong relationships and working together toward a common goal, stakeholders can also foster a sense of community and pride in their local area. This can ultimately lead to a more positive and welcoming environment for visitors.

With the rise of water sports, recreation, and tourism, it is crucial to ensure that these activities are accessible and affordable for all individuals. This includes domestic travellers of all income levels. We should offer discounts, promotions, and special packages to encourage local residents to explore their own backyard.

This will not only encourage economic growth within the community but also allow individuals who may not have the means to participate in these activities to experience them. In addition to providing accessibility and affordability, it is important to consider the impact of these activities on the environment.

With the increase in popularity of water sports and tourism, there is also a potential for harm to the surrounding ecosystems. Therefore, it is essential to implement sustainable practises and regulations to ensure the preservation of these natural resources for future generations to enjoy. This can also be achieved by educating both visitors and locals on the importance of responsible tourism and the role it plays in protecting these environments.

By prioritising accessibility, affordability, and sustainability in water sports, recreation, and tourism, communities can benefit in more ways than just economic growth. These activities can also promote a sense of pride and connection to the local environment, as well as foster a more inclusive and diverse community. It is important to continue to strive toward these goals to not only enhance the experience for travellers but also to create a more harmonious and sustainable relationship between humans and nature.

Seasonal and event-based offerings are a great way to attract domestic tourists throughout the year. By capitalising on seasonal trends and special events, you can provide unique and exciting experiences that will leave a lasting impression on your guests. This could include organising water sports competitions, beach festivals, cultural celebrations, and themed events tailored to specific interests. These offerings will not only draw in visitors but also enhance their overall experience, making them more likely to return in the future. One of the key benefits of these types of offerings is their ability to cater to different interests and preferences. For example, a beach festival may attract families with children, while a cultural

celebration may attract history buffs and art enthusiasts.

By diversifying your seasonal and event-based offerings, you can appeal to a wider audience and increase your chances of attracting more visitors. Plus, these events can also serve as a great way to showcase local culture and tradition, giving travellers a deeper understanding of and appreciation for the destination. In addition to attracting visitors, seasonal and event-based offerings can also help boost the local economy.

By organising these events, you can create job opportunities for the community and generate revenue for local businesses. This not only benefits the destination as a whole but also adds to the overall experience for travellers. So, don't miss out on the opportunity to capitalize on seasonal trends and special events to attract and engage domestic travellers throughout the year.

2: Requalification and Reassessment of Water Domain Areas

1. - Use of marginal riverside areas

Waterfront areas are often forgotten portions of any major city. However, these spaces can be revived and repurposed to serve a vital role in urban life. By converting and using marginal areas, cities can benefit from increased aesthetic appeal and a new source of income. We can transform abandoned or underutilized spaces into vibrant hubs for water-based recreation, like marinas, waterfront parks, and beachfront promenades.

This not only brings new life to these areas but also provides opportunities for both locals and tourists to engage in recreational activities.

The benefits of repurposing degraded ports and marginal areas extend beyond just economic gain. By creating new spaces for water sports and recreational activities, cities can also improve the overall well-being of their residents. These areas provide a respite from the hustle and bustle of city life, allowing people to connect with nature and engage in physical activities. Additionally, they can serve as a social hub, bringing people together and fostering a sense of community. By breathing new life into these forgotten spaces, cities can create a more liveable environment for their residents.

80

1. - Environmental recovery and conservation

Implementing regulations and guidelines for the orderly and disciplined use of coastal water plans is critical to maintaining the health of our oceans and marine life. These regulations provide a framework to support sustainable management of marine resources and minimise environmental degradation. By setting up zoning regulations, designated areas for specific activities, and enforcement measures, we can protect our coastal areas from overcrowding and pollution.

Zoning regulations help control the types of activities that can take place in different areas of the coast. For example, certain areas may be designated for fishing, while others may be reserved for recreational activities such as swimming and surfing. This not only ensures that the ecosystem is not disturbed by incompatible activities but also helps prevent conflicts between different user groups.

In addition, designated areas for specific activities can also help protect vulnerable species and habitats, preserving the diversity of marine life. Enforcement measures are also an important aspect of implementing regulations for coastal water plans. These measures ensure that the regulations are followed and help prevent illegal activities that may harm the environment. For instance, strict penalties may be enforced for polluting or damaging coastal areas. This not only acts as a deterrent for potential offenders but also sends a strong message about the importance of protecting our oceans. By implementing these regulations and guidelines, we can work toward more sustainable and environmentally friendly use of our coastal waters.

Water Domain Areas Requalification and Revaluation: Water domain areas, including rivers, estuaries, and coastal zones, play

81

critical roles in sustaining local ecosystems. However, many of these areas have been neglected and are currently facing threats such as pollution and habitat degradation. Thus, there is a growing need for the requalification and revaluation of these areas to improve their ecological value. By doing so, not only will these areas become more resilient against environmental challenges, but they will also provide numerous opportunities for recreational activities. One way to achieve a balance between ecological and recreational values is through habitat restoration projects. These projects involve the rehabilitation of degraded habitats, which can benefit a wide range of aquatic and terrestrial species.

Furthermore, water quality improvement initiatives can be implemented to reduce pollution levels and maintain a healthy ecosystem. This will not only benefit wildlife but also provide better conditions for recreational activities such as kayaking, paddle boarding, and wildlife watching. Moreover, enhancing public access to these areas can encourage more people to appreciate and protect these valuable natural resources. In addition to the ecological and recreational benefits, the requalification and revaluation of water domain areas can also have positive impacts on local economies.

Promoting and developing these recreational areas can attract tourists and generate revenue for the surrounding communities. Thus, it is not only important to protect and restore these areas for their ecological value but also for their potential economic value. With proper management and planning, the requalification and revaluation of water domain areas can bring about a more sustainable and prosperous future for both the environment and local communities.

Environmental Restoration and Conservation: In addition to mitigating negative impacts, integrating environmental restoration and conservation efforts can enhance the experience of water sports and recreational development projects. By preserving and restoring natural habitats, visitors can enjoy the beauty and diversity of the surrounding ecosystem. This can also create educational opportunities for visitors to learn about the importance of environmental conservation and how they can help protect these sensitive areas.

In addition, incorporating these measures can improve the overall sustainability of water sports and recreational development projects. By preserving and restoring natural habitats, these projects can reduce their carbon footprint and promote eco-friendly practises. This can also attract environmentally conscious visitors, boosting the project's reputation as a responsible and sustainable destination. Ultimately, integrating environmental restoration and conservation efforts into water sports and recreational development projects not only benefits the ecosystem but also enhances the visitor experience and promotes sustainability.

By working together with nature, these projects can create a harmonious balance between human activities and the environment. This approach not only helps protect sensitive ecosystems but also creates a more enjoyable and sustainable environment for everyone to enjoy.

1. - Maritime Transport, Ports and Logistics

The rapid expansion of world trade is a clear indicator of the ever-increasing globalisation of business operations. As trade barriers continue to be removed and technological advancements make it easier to connect with suppliers and customers around the world, companies are now able to operate on a global scale like never before.

This has led to an exponential increase in trade, surpassing even the growth rate of world production. With the rise of international trade, companies are now able to access a wider range of suppliers and customers, leading to more competitive prices and higher-quality products.

This has also created new opportunities for businesses to expand their operations to new markets, further fueling the growth of world trade. Furthermore, economic integration has also made it easier for companies to source raw materials and labour from different parts of the world, resulting in more efficient production processes and increased profitability. However, along with the benefits, the increasing globalisation of operations has also raised concerns about the impact on local economies and employment.

As companies move their operations to other countries, it can lead to the loss of jobs in the home country. Furthermore, reliance on global supply chains can make businesses vulnerable to disruptions and fluctuations in the global economy. Therefore, while the growth of world trade is a positive sign of a connected and interdependent world, it is important to carefully consider the implications and find ways to mitigate potential risks.

Global value chains, also known as GVCs, are complex processes that have been developed over the years in response to the growing need for efficiency and cost-effectiveness in the production of goods.

84

These chains involve various stages of production, from the initial design and development of a product to its final assembly and distribution. Each stage of production takes place in different countries, allowing firms to take advantage of the comparative advantages of each region. This has led to a significant increase in cross-border trade in intermediate goods.

One of the main advantages of GVCs is the ability to tap into the specialised skills and resources of different countries. For example, a company may design and develop a product in one country, source raw materials from another, and then assemble the final product in a third country. This allows firms to access the best inputs at the lowest cost, increasing their competitiveness in the global market. Furthermore, GVCs provide opportunities for developing countries to participate in global trade and benefit from international market growth. However, the fragmentation of production processes also brings challenges.

With different stages of production taking place in different countries, coordination and communication between firms have become crucial. This can be a complex and time-consuming process, as firms must navigate different legal, cultural, and logistical barriers. Additionally, the reliance on foreign inputs and foreign markets also exposes firms to risks, such as supply chain disruptions and changes in trade policies. As such, while GVCs offer many benefits, they also require careful management and adaptation to ensure their continued success.

Trade liberalisation has been a huge driving force in expanding international trade. As trade barriers are reduced and trade agreements are formed, the flow of goods and services across borders has become more seamless. The World Trade Organisation (WTO) agreements and regional trade pacts, like those with the European Union, have played a major role in this process.

Tariff reduction is one of the main benefits of trade liberalisation. This makes it easier and more affordable for countries to import and export goods to and from other nations. As a result, businesses have been able to expand their markets and reach a wider customer base. This has also led to increased competition, which can drive down prices and improve the quality of goods and services.

However, there are also concerns about the impact of trade liberalisation on developing countries. While it has opened up new markets and opportunities, it has also made it difficult for smaller, less developed economies to compete with larger, more advanced ones. This has led to calls for fair trade practises and domestic industry protection. Overall, trade liberalisation has had a significant impact on the global economy and played a crucial role in the growth of international trade.

Technological advancements are constantly evolving and have made considerable changes in global trade. The introduction of containerisation, air freight, and digital communication has revolutionised transportation and communication. This has made it easier and more affordable to transfer goods and information over long distances.

These technological advancements have led to an increase in the speed of global trade, making it more efficient and cost-effective. Containerisation, or the use of large containers to transport goods, has simplified the process of loading and unloading cargo. This reduces the time and labour requirements, making it more efficient and cost-effective. In addition, air freight has significantly reduced the time it takes to transport goods across long distances. This has allowed businesses to expand their reach and access new markets.

Moreover, digital communication technologies have also played a crucial role in the acceleration of global trade. The ability to instantly communicate and share information has made it easier for businesses to coordinate and manage international trade. This has also opened up new opportunities for businesses to expand their networks and access a wider range of suppliers and customers. Overall, these technological advancements have greatly contributed to the growth and globalisation of trade, making it easier and more efficient to move goods and information across the world.

Employing an external organization to carry out an in-house business function is known as outsourcing.

Companies frequently use this strategy to reduce labor costs or access specialized expertise that may not be available in their own country.

One form of outsourcing is offshoring, which involves relocating production or services to a country with lower labour costs or specialised knowledge.

Off shoring has become increasingly popular due to the benefits it offers companies. By relocating production or services to a country with lower labour costs, companies can save money on wages and other expenses. Furthermore, offshoring can provide access to specialized skills and resources that may not be available in the company's home country. This can lead to increased productivity and efficiency, as well as improved quality of goods and services.

However, outsourcing also has its drawbacks. While it may benefit companies, moving production and services overseas can also result in job losses in the home country. This can have a negative impact on the local economy and lead to social and political tensions. Additionally, off shoring can also pose challenges in terms of communication, cultural differences, and legal regulations. It is important for companies to carefully consider the potential risks and benefits before deciding to outsource or offshore their operations.

Container shipping has revolutionised the global trade industry, providing unmatched efficiency and scalability when it comes to transporting goods across vast distances. With standardised container formats, the process of transferring goods between different modes of transportation, including ships, trucks, and trains, has become seamless and hassle-free. This has significantly streamlined logistics operations, resulting in reduced costs and increased profitability for businesses worldwide.

The popularity of container shipping can be attributed to its unmatched efficiency and cost-effectiveness. The standardised containers make it easy to load and unload goods, minimising the chances of human error and damage to the cargo. As a result, container shipping has become the preferred mode of transportation for global trade, with a significant increase in demand over the years. Container shipping's growth has been further fuelled by the increasing globalization of trade, making it an essential part of the global economy. In addition to its efficiency and scalability, container shipping also offers unparalleled security for goods in transit. The standardized containers are equipped with advanced tracking systems that allow businesses to keep a close eye on their cargo's movement.

Maritime transport activities are commonly categorized into five distinct segments, each serving different types of cargo:

1. - Solid Bulk

Solid bulk transport is crucial to the global economy as it allows for the transportation of essential commodities such as cereals, soybeans, and minerals. These dry bulk materials are necessary for the growth of industries around the world, and without them, the economy would come to a halt. Bulk carriers are specifically designed to transport these goods, and they have large cargo holds that allow for the efficient loading and unloading of bulk commodities.

These cargo ships are an integral part of the shipping industry and play a major role in international trade. They transport goods across vast distances, making it possible for countries to import and export essential materials. These ships' size allows them to carry large amounts of cargo, reducing the cost of transportation per unit of material. This makes it possible for these dry bulk commodities to be accessible to consumers at a reasonable price. Transporting solid bulk materials is not without its challenges.

Ships must navigate through rough seas and unpredictable weather conditions, making their journey treacherous. Additionally, the loading and unloading of bulk carriers requires specialised equipment and skilled workers to ensure safe and efficient transport. However, despite these challenges, solid bulk transport remains a vital aspect of global trade, connecting countries and keeping economies running smoothly.

89

1. - Liquid bulk transport

Liquid bulk transport is an important sector of the shipping industry, as it handles the transportation of large quantities of liquid cargo. These cargoes can range from oil products (such as crude oil and refined petroleum) to chemicals and liquefied natural gas. Specialized vessels known as tankers transport these commodities due to their hazardous nature.

These tankers have onboard tanks that are specifically designed for storing and transporting liquids in a safe and efficient manner. The process of liquid bulk transport involves careful planning and coordination to ensure that the cargo is delivered to its destination without any incidents. This includes adhering to strict safety protocols and regulations, as well as using state-of-the-art technology to monitor and control the cargo's temperature and pressure. Tankers are also equipped with advanced navigation and communication systems to ensure smooth and efficient transport.

Aside from the technical aspects, liquid bulk transport also plays a crucial role in the global economy. It enables the efficient movement of vital resources, such as oil and gas, which are essential for various industries and everyday life. Moreover, the industry provides numerous job opportunities for skilled workers, including engineers, captains, and crew members. As the demand for liquid bulk transport continues to grow, the industry is constantly evolving and improving to meet the market's ever-changing needs.

90

1. - General cargo

General break-bulk cargo is the most common type of cargo transported by sea, accounting for a large portion of global trade. This type of cargo is made up of a wide variety of goods that are packaged separately rather than being loaded in bulk. These goods can vary greatly in size, shape, and weight, making them challenging to transport. This General cargo ships equip themselves with specialized cargo handling gear, such as cranes and forklifts, to efficiently load and unload the cargo.

Items that fall under the general break-bulk cargo category include heavy machinery and equipment, steel products, and even consumer goods.

These goods are usually loaded onto pallets, crates, or containers for easier handling and transportation. This method of shipping is preferred for goods that cannot be transported in bulk, such as liquid or granular materials. General cargo ships are designed to transport a wide variety of goods, making them versatile vessels that play a crucial role in global trade. One of the main advantages of general break-bulk cargo is its flexibility.

Various modes such as trucks, trains, and ships can transport general cargo, unlike bulk cargo, which necessitates specialized vessels and facilities for loading and unloading.

This makes it easier for businesses to transport their goods to different parts of the world, providing them with greater market access and opportunities for growth. However, handling and transporting general cargo also come with their own set of challenges, such as the need for careful packaging and proper securing of the cargo to prevent damage during transit.

91

1. - Roll-on/Roll-off (Ro-Ro) transport

Roll-on/Roll-off (Ro-Ro) transport is a highly efficient and specialised method of transporting wheeled cargo. : Cars, trucks, trailers, and other vehicles that can drive onto and off of the specialized Ro-Ro vessels are ideal for this type of transportation.These vessels feature ramps and decks that are specifically designed to make the loading and unloading process as quick and efficient as possible.

The Ro-Ro vessels are equipped with state-of-the-art technology to ensure the safe and secure transport of wheeled cargo. This includes features such as specialised lashing systems and adjustable decks to accommodate different types of cargo. This not only ensures the safety of the cargo but also makes the process of loading and unloading much smoother and more efficient.

One of the main advantages of Ro-Ro transport is its speed and reliability. With an efficient loading and unloading process, cargo can be transported quickly and with minimal delays. This makes it a preferred choice for time-sensitive shipments. Additionally, Ro-Ro transport offers cost savings because it eliminates the need for expensive cranes and other heavy machinery for loading and unloading. Overall, Ro-Ro transport is a vital mode of transportation for wheeled cargo and plays a crucial role in keeping the global supply chain moving smoothly.

92

1. - Containerised transport

Containerised transport has revolutionised the shipping industry by providing a standardised way to transport a diverse range of goods. This has been made possible by the use of containers that can be easily loaded onto container ships, which are designed with cellular holds that can accommodate containers of various sizes. This has made it easier to transport goods across long distances, as well as significantly reducing shipping time and cost.

The numerous benefits of containerization have driven its widespread adoption. These include enhanced efficiency, security, and scalability, making it the preferred mode of transport for various industries. This has fuelled the container shipping industry's rapid growth, which has been further accelerated by factors such as globalization, trade liberalization, and e-commerce expansion. As a result, containerised transport has become the dominant mode of international trade, especially for manufactured goods. Standard containers have also enabled economies of scale, making it more cost-effective to transport large quantities of goods. This has made containerised transport a popular choice for businesses, contributing to the growth of international trade.

Additionally, containerization enables multimodal transport capabilities, facilitating the easy transfer of goods between various modes of transportation like ships, trains, and trucks.

This has further increased the efficiency and flexibility of containerised transport, making it a crucial component of the global supply chain.

In recent years, LNG transportation has grown significantly. This is due to the increased demand for natural gas as a cleaner alternative to traditional fossil fuels. With countries around the world looking for ways to reduce their carbon footprint, natural gas has become a popular choice.

93

As a result, the use of specialised vessels, known as LNG carriers, has increased. These ships are equipped with cryogenic tanks that can store LNG at extremely low temperatures, making it easier to transport it from production facilities to markets. The expansion of LNG infrastructure has also played a crucial role in the growth of LNG transport. As more countries invest in the development of new liquefaction and degasification projects, the demand for LNG carriers has increased. This has created new job opportunities and boosted the global economy. Additionally, the use of LNG as a cleaner alternative to traditional fossil fuels has helped reduce air pollution and combat climate change.

Overall, LNG transportation is a vital part of the global energy industry. It has not only provided countries with a cleaner source of fuel but has also contributed to the growth of the global economy. With the continued expansion of LNG infrastructure and the development of new projects, the future of LNG transport looks promising.

1. - Deep-sea transport

Is a vital component in the exchange of resources between nations and the lifeline of international trade. It facilitates the movement of goods and commodities across vast distances, such as raw materials, consumer products, and industrial equipment, connecting businesses and consumers worldwide.

This highly complex process involves various stages, from packaging and loading to navigating through unpredictable weather conditions and customs regulations. It relies on a network of ports, shipping lines, and logistics providers to ensure the timely and cost-efficient delivery of goods.

The impact of deep-sea shipping on the global economy can be seen as significant. By providing access to markets and facilitating the exchange of goods, it contributes to industry growth and creates employment opportunities. It also allows countries to focus on their comparative advantages, such as natural resources and skilled labour, in order to participate in the global market. Additionally, deep-sea shipping allows goods to flow to regions that lack certain resources, promoting economic development and improving the standard of living. As a result, it plays a crucial role in promoting international relations and fostering global cooperation.

This kind of shipping is essential for international trade, allowing for the exchange of various products and resources between different regions. The ships used for deep-sea shipping are specifically designed to withstand the harsh conditions of the open ocean, making them ideal for long journeys.

The operations involved in deep-sea shipping are complex and require careful planning and coordination. From loading and unloading cargo to navigating through unpredictable weather conditions, every aspect of these voyages must be carefully managed. In addition, deep-sea shipping also involves complying with various

95

international laws and regulations, ensuring the safety of both the crew and the cargo. Despite these challenges, the benefits of deep-sea shipping are vast, enabling the global economy to thrive and grow. Deep-sea shipping's impact extends far beyond the transportation of goods and commodities. It also plays a significant role in shaping international relations and promoting cultural exchange. As ships travel between different countries and ports, they bring people and cultures together, creating connections and fostering understanding between nations. This makes deep-sea shipping not only an essential part of the global economy but also a vital element in promoting global cooperation and unity.

Deep-sea shipping vessels are often used to transport goods across oceans. They are specifically designed to handle the challenges of long-distance voyages and open-ocean navigation. These vessels have the ability to travel for extended periods of time and can carry large quantities of goods. This is why they are commonly used to transport cargo from one continent to another.

There are different types of deep-sea shipping vessels, each with their own unique features. Container ships are used to transport goods in standard shipping containers, making loading and unloading cargo easier. Bulk carriers are designed to transport large quantities of unpackaged goods such as coal, grains, and iron ore. Tankers, on the other hand, are used to transport liquids like oil and gas.

These vessels are equipped with special tanks that can hold thousands of gallons of liquid. Safety is of utmost importance in the shipping industry, especially when it comes to transporting goods across oceans. That's why deep-sea shipping vessels are equipped with advanced navigation systems and safety features. These vessels are also equipped with cargo handling equipment to ensure the efficient loading and unloading of goods. This not only saves time but also reduces the risk of damage to the cargo. With these features, deep-sea shipping vessels are able to transport goods safely and efficiently, making them an integral part of the global economy.

The global demand for deep-sea shipping continues to rise with the increasing need for international trade. This mode of transportation enables the transport of a diverse range of goods and commodities across the world. From the manufacturing hubs of one country to the consumer markets of another, deep-sea shipping plays a crucial role in sustaining the global economy.

Container ships have revolutionised the shipping industry by streamlining the transportation of manufactured goods and consumer products. These massive vessels are designed to carry thousands of standard shipping containers, making it easier to load and unload large quantities of goods efficiently. This has greatly reduced the time and cost of transportation, making it a popular choice for businesses worldwide.

However, we use specialized vessels like bulk carriers and tankers for bulk commodities like coal, iron ore, and grains. These ships are specifically designed to transport dry bulk and liquid cargoes, respectively, in large quantities. With advanced technology and efficient logistics, deep-sea shipping has become an essential component of global trade. This mode transports everything, from small items like mobile phones to large machinery. It has not only made international trade more accessible but also opened up new markets for businesses to explore. It plays a significant role in connecting people and economies, creating a more interconnected and interdependent world.

While deep-sea shipping has been in operation for centuries, it wasn't until the late 19th century that it began to take on its current form. With the advent of steam-powered vessels and the opening of the Suez Canal, long-distance shipping became much more efficient, allowing for the establishment of trade routes between major production centres and consumption markets around the world.

These trade routes, which are still in use today, include the Trans-Pacific route between Asia and North America, the Trans-Atlantic route between Europe and North America, and the Asia-Europe route linking manufacturing hubs in Asia with markets in Europe. These major shipping lanes have played a crucial role in shaping the global economy, allowing for the efficient transport of goods and raw materials across vast distances. Without these established trade routes, it would be much more difficult and costly for businesses to reach international markets, and the global economy would not be as interconnected as it is today. In fact, the growth and development of many industries, such as the automotive and electronics industries, would not have been possible without the reliable and efficient transportation provided by deep sea shipping along these trade routes.

Today, deep-sea shipping continues to play a vital role in the global economy, with thousands of cargo ships traversing these trade routes every day. However, as the demand for goods and resources continues to grow, there is also a growing concern for the environmental impact of these shipping operations. New technologies and alternative fuels are being developed to make deep-sea shipping more sustainable and environmentally friendly. Despite these challenges, deep-sea shipping remains an integral part of the global trade network and will continue to shape the global economy for years to come.

Deep-sea shipping is widely regarded as the backbone of international trade and a significant contributor to the global economy. It allows countries to access distant markets, providing them with essential resources and opening up opportunities for economic growth and development. The efficiency and scale of deep sea shipping allow for the transportation of large quantities of goods at a relatively low cost, making it an essential part of global connectivity.

Deep-sea shipping's reliable and cost-effective transportation solutions have revolutionized the way countries interact and trade. The ability to transport goods over long distances has allowed for the creation of global supply chains, where individual countries specialise in producing certain goods and rely on others for the products they need. This has led to increased efficiency and productivity, benefiting both producers and consumers.

Without deep-sea shipping, many countries would not be able to participate in international trade as effectively, limiting their economic potential. The reliable and cost-effective transportation solutions provided by this industry have played a crucial role in facilitating global connectivity and economic integration, allowing for the exchange of goods and resources across borders.

As long as global trade continues to expand, deep-sea shipping will remain an essential part of the global economy.

1. - Short Sea Shipping

Short-sea shipping, or SSS, is an environmentally friendly mode of transportation that offers a sustainable alternative to traditional land-based shipping methods. : People often use this type of shipping to transport goods and passengers over relatively short distances, like along coastlines or between neighboring countries.

It is a cost-effective means of transport that reduces road congestion and carbon emissions, making it an attractive option for businesses and governments alike. One of the primary advantages of SSS is its ability to ease the burden on crowded roadways. By utilising coastal routes, this mode of transportation reduces the number of trucks on the road, which in turn reduces traffic congestion and the associated air pollution.

Additionally, SSS provides a more efficient way to transport goods and passengers, as ships have a larger capacity and can carry heavier loads than trucks. Furthermore, SSS promotes regional trade and economic growth by connecting ports and cities within a single region. This allows for the efficient and timely delivery of goods and services, which in turn supports local businesses and boosts the economy. Moreover, SSS provides a vital link for trade between neighbouring countries, fosters economic cooperation, and strengthens diplomatic relations. Because of these advantages, short-sea shipping is an important component of global trade and transportation networks.

Short-sea shipping involves the transportation of goods and passengers through waterways, such as rivers, lakes, and canals, which are located close to the shore. Unlike deep sea shipping, which primarily caters to international trade, short sea shipping is more focused on domestic trade, where the distances covered are not very long.

100

Ports and harbors along the coast, inland areas, and other regions close to the sea frequently use this mode of transport.

One of the main advantages of short-sea shipping is that it reduces the reliance on road and rail transportation, which can be costly and time-consuming. This shipping method also helps in reducing carbon emissions and road congestion, as ships can carry a much larger volume of goods compared to trucks and trains.

Moreover, short-sea shipping provides a more environmentally friendly alternative to long-haul trucking, as ships use less fuel and produce fewer emissions. This makes it a popular choice for transporting goods that are not time-sensitive and can be delivered over a longer period of time. Another benefit of short-sea shipping is that it provides a more cost-effective and efficient way of transporting goods to remote and poorly connected areas. This is especially useful in countries with long coastlines, numerous islands, and a vast network of rivers and canals. It also helps boost local economies and create job opportunities in these areas. Additionally, short-sea shipping can offer a more reliable and flexible service compared to other modes of transportation, as it is less affected by weather conditions and traffic congestion. Overall, short-sea shipping plays a crucial role in promoting sustainable and efficient trade within a country's borders.

Short-sea shipping is an essential aspect of the transportation industry, playing a crucial role in connecting various ports, terminals, and communities along coastlines. This mode of transportation provides efficient and reliable links within a specific geographic region or maritime area. It allows for easy intra-regional trade and transportation, thereby promoting economic activity in the region.

One of the major advantages of short-sea shipping is its ability to reduce road congestion, which is a significant problem in many coastal regions. With the increasing number of vehicles on the roads, short-sea shipping offers a sustainable alternative for transporting goods and people. It also helps to reduce carbon emissions, making it an environmentally friendly mode of transportation.

Moreover, short-sea shipping is a cost-effective method of transporting goods, especially over short distances. It is relatively cheaper than other modes of transportation, such as road or air transport. This makes it an attractive option for businesses and industries looking to transport goods within a specific region. Additionally, it provides a reliable and timely service, ensuring that goods reach their destination on time. Overall, short-sea shipping is an important aspect of the transportation industry, contributing to the growth and development of coastal regions.

Short-shipping is an essential part of the transportation industry, providing a crucial link between producers and consumers. It not only transports goods and commodities such as containers, bulk cargo, and general cargo but also plays a significant role in promoting local and regional economies.

For instance, it facilitates the movement of agricultural products, ensuring that fresh produce reaches markets in a timely and efficient manner. Additionally, short-sea shipping supports sectors like manufacturing and construction by transporting raw materials and finished goods to and from their respective destinations. Moreover, short-sea shipping is a vital component of the tourism industry, connecting coastal regions and creating opportunities for leisure and recreation. It enables passengers to access popular tourist destinations, enhancing the overall travel experience.

Furthermore, short-sea shipping plays a crucial role in supporting retail industries by ensuring the timely delivery of goods to meet consumer demands. Its versatility in transporting a wide variety of goods and commodities makes it a vital link in the supply chain, meeting the diverse transportation needs of different markets. In conclusion, short-sea shipping is a crucial mode of transportation that serves multiple purposes and sectors. Its ability to transport various goods and commodities, including passengers, makes it an integral part of the global economy.

Short-sea shipping not only facilitates trade and commerce but also promotes the growth of local and regional markets. It is a sustainable and efficient transportation option that plays a crucial role in meeting the diverse needs of modern society.

It is a highly efficient mode of transportation that offers shippers and customers flexibility and timely delivery options. With frequent sailings and flexible schedules, short-sea shipping services are able to provide tailored transportation solutions to meet specific needs. These services are known for their reliability and responsive nature, making them a popular choice for many businesses. Furthermore, short-sea shipping is characterised by its cost effectiveness and environmental friendliness. By utilising waterways, this mode of transportation reduces the carbon footprint and helps alleviate traffic congestion on the roads. Furthermore, regular sailings and schedule flexibility allow for better resource planning and optimisation, resulting in cost savings for shippers and customers. Overall, short-sea shipping services are an essential component of the global transportation industry. They offer a wide range of benefits to businesses and individuals, including efficiency, reliability, cost-effectiveness, and environmental sustainability.

Many businesses use it as a major component of global trade to move goods from one country to another. It is also a popular mode of transportation for people who want to travel between countries without taking a flight.

This is because it is frequently viewed as a more environmentally friendly option compared to other transport methods. It provides a number of benefits, such as reducing greenhouse gas emissions, easing congestion on roadways, and consuming less fuel in comparison to land-based transport. One of the main reasons why short-sea shipping is considered environmentally friendly is because it helps to mitigate the negative impacts associated with overland transportation.

As goods and people are moved through the sea, it reduces the need for trucks and other vehicles to travel long distances on land. This, in turn, results in a decrease in carbon emissions, which helps combat climate change. In addition, short-sea shipping contributes to sustainable mobility and logistics by providing a more cost-effective and efficient alternative to traditional transportation methods. However, despite its many advantages, short-sea shipping still faces challenges and limitations. For instance, it is not always a viable option for every type of cargo or for certain destinations. There may also be concerns about port infrastructure and the availability of suitable vessels.

Nevertheless, with advancements in technology and a growing focus on sustainability, it is likely that short-sea shipping will continue to play a significant role in global transportation and trade, offering both economic and environmental benefits.

Short-sea shipping initiatives are designed to reduce road and rail transport by promoting the use of coastal and inland waterways. This is beneficial for both the environment and the economy. By shifting the transport of certain types of cargo and passengers from road and rail to maritime, we can alleviate road congestion and improve air quality. Waterways, often underutilized compared to roads and railways, also contribute to enhanced transport efficiency. Furthermore, short-sea shipping has numerous other advantages.

Not only does it promote the use of environmentally friendly transport methods, but it also supports the development of coastal and inland regions. In these areas, waterways create opportunities for economic growth and job creation. In addition, short-sea shipping can help reduce the transport sector's carbon footprint, making it a crucial tool in the fight against climate change. In conclusion, short-sea shipping initiatives have a significant impact on both the environment and the economy.

By shifting certain types of cargo and passenger movements from road and rail to maritime transport, we can alleviate road congestion, improve air quality, and enhance transport efficiency. This also leads to economic benefits, such as job creation and reduced carbon emissions. Short-sea shipping is a critical component of sustainable transportation that should be promoted and supported for a better future.

1. – Feedering

Feedering is a logistical concept that involves transporting cargo between a main port and smaller secondary ports using smaller vessels, thus connecting these secondary ports to the larger global shipping network.

This concept is crucial for the efficient functioning of international trade and commerce. It allows goods to be transported to and from smaller ports that may not have the necessary infrastructure to handle large cargo vessels. The smaller vessels used for feeding are typically more efficient and can easily navigate the shallower waters of secondary ports. Feedering has been a game-changer in the shipping industry, especially for developing countries with smaller ports.

It has contributed significantly to the growth of these countries' economies by providing an affordable and efficient means of transporting goods. Smaller ports can now be connected to larger shipping networks via feedering, allowing them to import and export goods to and from various parts of the world. Without this concept, these ports would have remained isolated, limiting their potential for growth and development.

Additionally, feeding has had a positive impact on the environment Smaller vessels have reduced the transportation's carbon footprint, leading to cleaner and greener shipping practices.

It has also reduced the need for large cargo vessels to enter smaller ports, which can cause significant damage to the ecosystem. Therefore, feedering not only benefits the economy but also promotes sustainable and eco-friendly practises in the shipping industry.

It plays a vital role in the shipping industry by providing a seamless distribution mechanism for cargo arriving at major hub ports. This process of transferring cargo from large ocean-going

106

vessels at the main port to smaller feeder vessels helps to streamline the shipping process and ensures timely delivery of goods to secondary ports in the region. The integration of deep-sea shipping and feedering has transformed the global transportation of goods.

The use of feedering has greatly improved efficiency in the shipping industry, as it allows for the distribution of cargo to smaller ports that are not accessible by large vessels. This has opened up new trade routes and boosted the economies of remote regions that were previously unable to receive goods via sea transport.

Additionally, it has also helped to reduce congestion at major hub ports, allowing for smoother and faster movement of cargo.

By utilizing smaller feeder vessels for distribution, the shipping industry's carbon footprint has been reduced, as these vessels emit less pollution than their larger counterparts. The use of feeders has also reduced the need for trucks to transport goods from the main port to secondary ports, further contributing to a greener and more sustainable shipping industry.

It allows smaller ports to access essential goods and services that large container ships or other deep-sea vessels cannot directly provide. This cargo redistribution also helps alleviate congestion at busy main ports, ensuring timely delivery of goods and reducing overall shipping costs. Moreover, feeder vessels are crucial for developing countries that may not have the resources to build large ports capable of handling massive container ships. These countries can still participate in global trade by utilising feeder services, resulting in economic growth and development. This makes feeder vessels a vital part of the global supply chain, connecting businesses and consumers around the world. In addition to their economic benefits, feeder services also play a crucial role in disaster relief efforts. In times of crisis, feeder vessels can quickly transport essential supplies and aid to smaller ports that may have been affected by natural disasters or conflicts.

Feeder vessels are the backbone of maritime transportation, responsible for ferrying cargo from the main port to secondary ports and vice versa. These activities are strategically planned and executed within a network of ports, with the main port acting as the central hub and the surrounding secondary ports forming a web of interconnected routes. This network enables the efficient transfer of goods, expands the reach of maritime transportation, and connects various regions.

The feeder vessel system is crucial in maintaining a constant flow of goods and commodities between different parts of the world. It serves as a vital link in the supply chain, ensuring that cargo reaches its final destination in a timely and cost-effective manner. Feeder vessels are specifically designed to navigate through smaller ports and waterways, making them an essential component in connecting remote regions to the global market.

Feeder activities' success heavily depends on effective coordination and communication among all ports involved. Each port plays a critical role in the overall network, and any disruptions or delays at one port can have a ripple effect on the entire system. As a result, it is critical to carefully plan and manage feeder activities to ensure the network's smooth operation and continuous flow of goods.

There are many advantages to this method, one of the most important being improved accessibility to global markets for smaller ports. This means that even ports that may not have been able to handle large ships in the past are now able to receive goods from all over the world. This opens up many opportunities for these smaller ports, allowing them to compete with larger ports and expand their businesses.

Another major benefit of feedering is the reduction of congestion and vessel waiting times at main ports. This is a significant issue for many ports, as long wait times can lead to delays and increased costs for both shippers and carriers.

Feedering reduces traffic and congestion at larger ports by transporting cargo to smaller ports and then distributing it to main ports.

This, results in a smoother and more efficient transportation process for all parties involved. It provides enhanced flexibility in cargo distribution. This means that shippers in regions served by secondary ports now have access to cost-effective transportation solutions. This is especially beneficial for businesses that may not have the resources to transport goods directly to main ports.

Feedering is an essential aspect of the maritime industry, as it allows the transportation of a wide variety of cargo. This encompasses containers, breakbulk cargo, and bulk commodities, among others.

Containers are particularly well-suited for feeding operations due to their standardised nature and ease of handling. They are also designed to be easily transferred between vessels. One of the main advantages is its ability to transport containers.

This means that they can be transported from one location to another without having to be unpacked and repacked, saving time and effort. : Additionally, the standardized size of containers simplifies their stacking and transportation.

Its ability to transport break-bulk cargo is another benefit. Ships load individual goods, not containerized, as break-bulk cargo.

This type of cargo includes machinery, vehicles, and other large and bulky items. Due to their irregular shape and size, break-bulk cargo can be challenging to transport.

It is a vital link that connects secondary ports to main trade routes and distribution networks. It serves as a conduit for the movement of goods and commodities within a particular region, promoting economic integration and enhancing the competitiveness of local industries and markets.

By connecting smaller ports to larger ones, it reduces the dependency on a single transportation mode, making the trade network more robust and less susceptible to disruptions.

1. - Cabotage

Cabotage is a term used to describe the transport of passengers, cargo, and goods within a single country's coastal waters and territorial boundaries.

It is a vital aspect of maritime operations, enabling the safe and efficient movement of people and goods between ports along the coastline or inland waterways. Cabotage is a crucial part of a country's economy, as it ensures the smooth flow of trade and commerce within its own borders.

One of the key benefits of cabotage is that it promotes the development and growth of domestic shipping industries. By limiting maritime transport operations to a single country, cabotage regulations encourage the use of local vessels and sailors, creating jobs and boosting the economy. It also allows for greater control and oversight of maritime activities, ensuring compliance with safety and environmental regulations. This, in turn, promotes the protection of the country's coastal waters and marine life.

However, cabotage regulations can also present challenges for international shipping companies, as they restrict their ability to operate within a country's waters. This can lead to higher costs and longer transit times for goods and passengers, as companies must adhere to strict regulations and use local vessels. Despite these challenges, cabotage remains an essential aspect of maritime transport, serving as a means of promoting domestic growth and protecting a country's maritime interests.

Cabotage operations are subject to specific regulations and legal frameworks established by national authorities to govern maritime transport activities within their jurisdiction. National authorities may include requirements related to vessel registration, crew nationality, cargo rights, and the types of cargo permitted for transportation in these regulations.

111

One of the main purposes of these regulations is to ensure the safety and security of all maritime activities within a country's borders. This includes the safety of the crew members, vessels, and cargo being transported. In addition to safety, these regulations also serve to protect the economic interests of a country.

By limiting cabotage rights to domestic vessels and crew members, national authorities are able to promote the growth of their own maritime industry and protect it from foreign competition. This ensures that the country's economy benefits from the transportation of goods and people within its own borders. Furthermore, these regulations also play a crucial role in maintaining national security. By controlling the types of cargo that can be transported within a country, authorities are able to monitor and prevent the illegal transportation of goods that may pose a threat to the country's security. This includes weapons, hazardous materials, and other prohibited items.

Cabotage activities are often used as a way to protect domestic shipping companies and their workers. By requiring vessels to be registered under the flag of the country in which they operate, governments can ensure that their own companies have access to the domestic market.

Foreign vessels cannot engage in cabotage activities unless they meet specific requirements and obtain the necessary licenses.

These regulations help maintain a balance between the national and international shipping industries.

In addition to protecting domestic shipping companies, cabotage activities also play a crucial role in ensuring the safety and security of a country's waters. By requiring vessels to be registered under the national flag, governments can hold them accountable for any incidents or violations that may occur while operating in domestic waters.

113

This helps to maintain a higher level of safety for both goods and passengers being transported between domestic ports. It also allows for the easier regulation and enforcement of environmental and labour laws, ensuring that all vessels operating in domestic waters follow the same standards.

The importance of Cabotage cannot be overstated when it comes to supporting domestic trade, commerce, and economic activities. This is accomplished by providing efficient and reliable maritime transportation services that cater to the needs of the domestic market. Apart from supporting the economy, cabotage also plays a crucial role in ensuring national security. This is achieved by regulating the movement of vessels in the country's territorial waters, thus preventing the entry of unauthorized vessels that may pose a threat to national security. Additionally, Cabotage allows for the efficient transportation of essential goods and supplies within the country, which is crucial during times of crisis or emergencies. It is an important aspect of maritime transportation that has far-reaching effects on the economy and national security. Its role in supporting domestic trade and commerce cannot be ignored, as it provides a reliable means of transportation for essential goods and supplies. Furthermore, it's contribution to national security by regulating the movement of vessels is vital in safeguarding the country's territorial waters.

Regulations are a set of rules that prevent foreign-flagged vessels from participating in domestic maritime transport. These rules were put in place to protect the interests of local vessels and ensure the safety of the domestic maritime industry. However, there may be certain exceptions or provisions under international agreements that allow for limited cabotage rights for foreign vessels in certain circumstances.

These exceptions have been put in place to facilitate trade and maritime cooperation between countries. These exceptions are often negotiated bilaterally or multilaterally between countries. This means that the countries involved come to an agreement that benefits both parties. By allowing limited cabotage rights, countries can open up opportunities for foreign vessels to engage in domestic maritime transport, which can help boost the economy. Furthermore, these agreements promote cooperation between countries, which can lead to better relationships and partnerships in the future.

1. – Ports

The ports, once small and bustling with fishermen and small ships, have now been transformed into "mega hubs". These known as major transhipment centres, are home to large operators. The demand for speed has played a significant role in this transformation, with the need to move goods quickly and efficiently becoming the top priority. This shift has not only altered the physical appearance of these ports but also their organisation and regional role.

With the concentration of large operators in a select number of ports, competition has increased, and these ports have become more efficient and streamlined. This has led to an increase in trade and economic growth in the surrounding regions, making these crucial for global commerce.

However, this transformation has also brought about challenges and potential issues. The concentration of large operators in a few ports means that these ports are now under immense pressure to perform and handle large volumes of goods. This can lead to issues such as congestion, delays, and potential risks to the environment. Thus, while the shift toward megahubs has had major impacts, it is important for these ports to find a balance between efficiency and sustainability to ensure long-term success.

Megahubs are considered the lifeline of global trade because they play a critical role in facilitating the movement of goods and materials across different countries and continents. These hubs are known for their massive size and capacity, allowing them to handle a huge volume of containers and cargo every day.

This has led to massive investments in infrastructure and state-of-the-art technologies to improve efficiency and productivity. Many ports have invested in automated container terminals, which use advanced robotics and computer systems to handle and transport containers, reducing the need for manual labour.

115

Mega hubs, also known as mega ports, are essential platforms for both regional and global supply chains. These hubs are crucial for the consolidation, transhipment, and distribution of cargo across various regions and continents. They are crucial facilitators for the movement of goods between production centres and consumption markets, acting as trade gateways. Their strategic location and function make them key players in international trade.

This growth and investment are not limited to the ports themselves but also extend to the surrounding areas, creating a ripple effect of economic prosperity Megahubs play a critical role in connecting different regions and continents through trade, underscoring their importance.

They serve as key players in the global economy, facilitating the movement of goods and driving economic growth. These hubs are essential strategic assets for their regions, and their significance will only continue to grow in the ever-evolving landscape of international trade.

With the rise of megahubs, the competition between transport companies to gain a strong foothold in the shipping industry has reached new heights. These hubs, bustling with activity, are home to large shipping lines and logistics providers, making them a prime location for businesses. As a result, ports are now in direct competition with each other to attract these key players by offering competitive tariffs, efficient services, and top-notch infrastructure.

The intense competition between these ports has been beneficial for both businesses and consumers. Companies now have the opportunity to choose from a variety of ports, each offering unique advantages and services. This has led to the improvement of port facilities and services, making them more efficient and cost-effective. With state-of-the-art infrastructure, ports are now able to handle larger volumes of cargo and provide faster turnaround times, ultimately benefiting the end consumer. In this cutthroat competition, ports must constantly innovate and adapt to attract and retain their customers. This has led to the development of specialised services, such as cold storage facilities and container tracking systems, to meet the specific needs of different industries.

With ports constantly striving to improve their offerings, the future looks promising for the shipping industry as a whole.

Mega hubs are more than just simple hubs; they are dynamic hubs designed to serve a variety of purposes. They are not just a place for goods to be stored and moved, but they are also equipped with a wide range of facilities that are designed to enhance the efficiency of the process. These hubs are designed to connect various activities, such as freight forwarding, warehousing, and distribution.

They also offer value-added logistics and other ancillary services. These services are designed to support the efficient movement and storage of goods, which further increases their attractiveness as transshipment centres. This means that not only are these hubs designed to facilitate the movement of goods, but they also offer various services that make the process smoother and more efficient. In addition to the services and facilities offered, megahubs also serve as a central hub for various activities. They are not just a place for goods to be stored and moved, but they also act as a hub for connecting different activities. This makes them crucial hubs in the logistics and supply chain industries. Megahubs, with their wide range of services and facilities, are truly dynamic hubs that play a vital role in the efficient movement and storage of goods.

1. - Port Security

Fencing is a crucial component of physical security measures at ports and terminals. In addition to preventing unauthorised access, fencing also serves as a deterrent to potential intruders. Access control systems are another important aspect of port security. Such systems use electronic devices, such as key cards or biometric scanners, to restrict entry to authorised personnel only. Security gates and barriers are also used to control the movement of people and vehicles in and out of the port. Surveillance is another important component of physical security at ports and terminals. Closed-circuit television (CCTV) cameras are strategically placed throughout the port to monitor and record activities. Motion sensors are also commonly used to detect any movement in restricted areas. Unauthorized access or suspicious activities trigger alarms, prompting security personnel to take immediate action.

These surveillance measures are crucial to ensuring the safety and security of the port and its operations.

It helps to protect the port's security by preventing unauthorized individuals from gaining access to restricted areas. With the many employees, visitors, and contractors that frequent ports every day, it is crucial to have screening procedures in place to ensure the safety of everyone involved. These screening procedures may include identity verification, background checks, and screening for prohibited items through metal detectors or X-ray scanners.

These measures help to identify potential threats and prevent them from entering the port, protecting the people and cargo within. It also helps to detect any illegal items that may be attempting to enter or leave the port, such as drugs or weapons.

In addition to these measures, port security personnel are trained to handle any situations that may arise during the screening process. This includes dealing with false positives or false alarms, as well as

118

handling individuals who may become agitated or uncooperative during the screening. With proper training and procedures in place, personnel screening plays a crucial role in maintaining the safety and security of ports around the world.

Vessel security is a crucial aspect of protecting our ports and waterways. In order to ensure the safety of the vessels docked at ports, measures have been implemented to enhance security protocols.

These measures include thorough screening of crew members, inspections of cargo, and monitoring of vessel movements. This comprehensive approach is designed to identify and prevent potential threats to the ships and their crew members.

To further strengthen the security of vessels, security personnel are often deployed to conduct patrols, inspections, and searches. These actions serve as deterrents to potential stowaways, smugglers, and other illicit activities that may occur onboard ships.

Regularly monitoring and inspecting the vessels allows security personnel to quickly identify and address any security concerns that may arise. This proactive approach helps minimise the risk of security breaches and ensures the safety of the port and its surroundings.

In addition to these measures, it is also important for vessels to have proper security plans in place. These plans should outline procedures for responding to security threats and provide guidelines for the crew members to follow in case of an emergency. By taking a comprehensive and proactive approach to vessel security, we can ensure the safety and security of our ports and vessels, as well as protect our vital waterways from potential threats.

1. - Security in Port Traffic

Cargo security is a critical aspect of port security systems, with procedures in place to inspect, screen, and track cargo throughout the supply chain. Advanced technologies such as radiation detectors, imaging systems, and electronic seals are used to identify and mitigate potential security threats posed by hazardous materials, contraband, or weapons concealed within cargo shipments.

This ensures that the cargo being transported is safe for the environment and the public, as well as preventing illegal activities from occurring. With the increasing number of cargo shipments being transported every day, it is crucial to have strict and efficient cargo security measures in place. This not only protects the cargo itself but also ensures the safety of the port workers and the surrounding community.

By utilising advanced technologies, authorities are able to quickly and thoroughly scan through all types of cargo, including hazardous materials, to detect any potential threats or illegal items. This not only saves time and resources but also reduces the risk of human error. In addition to the use of advanced technologies, cargo security also involves thorough inspection and tracking procedures.

This means that all cargo shipments are carefully screened and monitored from the point of origin to the final destination. This helps to prevent any unauthorised access or tampering with the cargo during transit. By having these procedures in place, cargo security systems are able to effectively mitigate potential security threats and ensure the safe and secure transportation of goods.

Ports are constantly under threat from cyber attacks. Hackers are targeting critical infrastructure and information systems, leaving ports vulnerable to ransomware and malware attacks. These attacks can cause major disruptions in communication and data flow.

To combat these threats, port security systems must include a variety of cyber security measures. These measures are designed to protect networks, communication systems, and data from any type of cyber intrusion.

This is crucial, as ports are responsible for handling a large amount of sensitive information, and any breach can have serious consequences. With the rise of technology and the increasing reliance on digital systems in ports, cyber security has become an essential aspect of port security. It is no longer enough to focus solely on physical security measures, as cyber attacks can cause just as much damage.

As such, ports must constantly update and improve their cyber security protocols to stay ahead of evolving threats. This ensures that ports can continue to operate efficiently and securely, without any disruptions or compromises to their systems.

Port security systems are critical tools for ensuring the safety of our ports and the constant flow of goods and services they provide. They are designed to provide a comprehensive response to a wide range of potential threats, including security incidents, natural disasters, and other emergencies.

This includes everything from coordinating with law enforcement agencies to working with emergency services and other port stakeholders to mitigate risks, manage crises, and ensure the continuity of port operations. The importance of port security systems cannot be overstated.

Without these systems in place, our ports would be vulnerable to a wide range of threats, both internal and external. These systems are designed to provide a robust defence against potential threats, allowing port authorities to quickly and effectively respond to any situation that may arise.

In addition, they help to ensure that the flow of goods and services remains uninterrupted, even in the face of unexpected emergencies or other disruptions. In order to be effective, port security systems must be constantly updated and maintained. This involves regular training and drills to test their effectiveness, as well as close coordination with all relevant stakeholders.

By working together, port authorities and their partners can ensure that our ports remain safe and secure and that they continue to play a vital role in our economy and daily lives.

Port security systems, also known as port security patrols, are an essential aspect of maritime security. These systems are designed to protect ports from potential threats by complying with international regulations and standards.

The International Maritime Organization (IMO) established the International Ship and Port Facility Security (ISPS) Code, one of these crucial regulations.

This code outlines the necessary security measures and protocols that ports must adhere to in order to maintain effective security. Compliance with these regulations is vital to ensuring the safety and security of ports and their surrounding areas. Without proper security measures in place, ports are vulnerable to various maritime security threats, such as terrorism, smuggling, and illegal immigration.

By complying with these standards, ports can better protect themselves and their neighbours communities from potential risks. In addition to complying with international regulations, port security systems must also regularly undergo evaluations and audits to ensure their effectiveness. These evaluations help identify any potential weaknesses or gaps in security measures and allow for necessary improvements to be made. By constantly reviewing and updating security protocols, ports can stay ahead of potential threats and maintain a high level of security.

A ship traffic surveillance system is a critical tool for ensuring maritime traffic safety and efficiency. It is a complex network that involves various technologies, procedures, and personnel working together to monitor and manage vessels within a specific area. This could include busy ports, harbours, or coastal zones with a high volume of ship traffic. A ship traffic surveillance system's primary goal is to prevent collisions between vessels.

With the use of advanced technologies such as radar, sonar, and GPS, operators are able to track the movements of ships in real-time and identify potential collision risks. This information is then communicated to the vessels, allowing them to take the necessary actions to avoid accidents.

Aside from collision prevention, a ship traffic surveillance system also plays a vital role in ensuring the efficient and orderly movement of vessels. With the help of trained personnel, the system can monitor and manage the flow of ships, especially in busy ports where there may be multiple vessels entering and leaving at the same time.

This helps to avoid congestion and delays, making the process of docking and unloading cargo more efficient. Overall, a ship traffic surveillance system is an essential tool for promoting safety and efficiency in the maritime industry.

Ship traffic surveillance systems are critical for maritime safety and security. These systems have now become more sophisticated with the use of radar and automatic identification system (AIS) technology. With these advancements, ships within a monitored area can be easily tracked using radar and AIS transponders.

This information is critical in determining vessels' positions, movements, and identities, as well as providing real-time information on vessel positions, headings, and speeds. Radar technology has been around for decades and is a crucial tool for ship traffic surveillance. It utilises radio waves to detect and track objects, including ships, within a certain range.

With the advancement of AIS technology, radar systems now have a more accurate and reliable source of information. AIS transponders, which are installed on ships, broadcast vessel data, including identification, position, course, and speed, providing a more comprehensive picture of the ship's movements.

Various traffic management measures are used to manage and control the flow of traffic within a specific area. This includes implementing traffic separation schemes, imposing speed restrictions, and providing navigational guidance to vessels. By using these measures, operators can ensure the smooth and safe flow of ship traffic.

One of the key features of these surveillance systems is the ability for operators to communicate with vessels in real-time. This is often done through radio or electronic means, allowing operators to provide navigational assistance or issue advisories to vessels. This ensures that captains and crew members are aware of any potential hazards or changes in traffic patterns, helping them make informed decisions to ensure safe navigation. Overall, ship traffic surveillance systems play a crucial role in ensuring the safety and efficiency of vessel movements.

They provide operators with the tools and information they need to effectively manage and control ship traffic, minimising the risk of accidents and improving overall maritime operations. These systems are continuously evolving and improving with technological advancements, making them an essential component of modern maritime traffic management.

One of the main objectives of a ship traffic surveillance system is to ensure that collisions between vessels are prevented. This is achieved by continuously monitoring vessel movements, which enables operators to identify any potential collision risks.

They are able to quickly and effectively respond to these risks by providing navigational assistance, issuing traffic advisories, and alerting vessels to avoid any dangerous situations. This is especially important in busy or congested areas, where vessels may have limited visibility of each other.

Through proactive measures, such as monitoring vessel movements and identifying potential risks, operators can effectively maintain safe separation distances between vessels. This is critical to ensuring the safety of all vessels operating in the monitored area. By continuously monitoring vessel movements, operators are able to stay on top of any changes or potential risks and take appropriate action to prevent collisions.

When an emergency arises on the water, the stakes are high. In the midst of a maritime incident or distress situation, the difference between a successful rescue and a tragedy can come down to mere minutes.

That's where ship traffic surveillance systems come in. By providing real-time tracking and coordination capabilities, these systems enable operators to quickly identify the location of vessels in distress. In addition to providing vital information on the location of vessels, these systems also allow for the efficient coordination of response efforts. This means that rescue resources can be dispatched to the scene quickly and effectively, potentially saving precious time and lives.

The ability to monitor ship traffic in real-time is a crucial aspect of emergency response and search and rescue operations, and these systems are a vital tool in ensuring the safety of those on the water. Ship traffic surveillance systems are important not only for emergency situations but also for ongoing monitoring and maintenance of maritime operations.

By continuously tracking vessel movements, operators can identify any potential issues or hazards and take proactive measures to prevent accidents or incidents from occurring. This not only ensures the safety of those at sea but also helps protect the environment and maintain the efficiency of maritime operations.

1. - Logistics Platforms

Logistics platforms play a crucial role in the distribution chain. They serve as more than mere links; they function as critical nodes of connection, convergence, and redistribution. These platforms are the backbone of an international network, allowing goods and services to flow smoothly from one location to another.

The efficiency of these platforms is critical to the success of any business or organisation. Logistics platforms are the key to connecting suppliers, manufacturers, and customers from all over the world. They serve as a bridge between different countries and continents, allowing for the exchange of goods and services. Without these platforms, the global economy would not function efficiently.

They are responsible for ensuring that products are delivered on time, in the right condition, and at the most cost-effective price. Moreover, logistics platforms act as critical nodes of convergence. They bring together various transportation modes, such as air, sea, and land, to facilitate the movement of goods. This convergence allows for a more streamlined and efficient flow of products, reducing transit times and costs.

They also serve as hubs for the redistribution of goods, allowing for shipment consolidation and route optimisation. In summary, logistics platforms play a vital role in the global supply chain, connecting businesses and consumers worldwide and ensuring the smooth flow of goods and services.

Logistics platforms play a crucial role in connecting various points within the supply chain. These platforms serve as the backbone of the supply chain, connecting production centres, transportation hubs, distribution centres, and end consumers.

They act as central points where goods flow in and out; they seamlessly link different stages of the supply chain. This efficient

126

coordination allows for the smooth and timely delivery of goods to the end consumer. Moreover, logistics platforms provide a more complete picture of the supply chain, allowing for better decision-making and optimisation. By utilising data and analytics, these platforms can identify inefficiencies and bottlenecks in the supply chain, providing insight into areas for improvement. This not only increases the supply chain's efficiency and effectiveness, but also reduces costs and improves customer satisfaction.

Despite their importance, logistics platforms must constantly evolve to keep up with the ever-changing demands of the supply chain. With the rise of e-commerce and the increasing demand for faster delivery times, these platforms must adapt to new technologies and innovations to remain competitive. This continuous evolution ensures that the supply chain remains efficient and effective, meeting the needs of both businesses and consumers.

Logistics platforms provide a centralised space that allows different parties involved in the supply chain to come together and work collaboratively. These parties include suppliers, manufacturers, carriers, distributors, and retailers.

At its core, a logistics platform is a hub for the exchange of goods, information, and resources. This means that all the stakeholders involved in the supply chain can easily share and access important data and resources. This fosters efficient communication and streamlines the supply chain process. Parties involved can make informed decisions that benefit the entire supply chain. In addition to acting as a central hub for collaboration and information exchange, logistics platforms also facilitate efficient supply chain management and operations. This implies that we can optimize processes like inventory management, order fulfillment, and transportation for optimal efficiency.

Logistics platforms are an essential part of modern-day global trade. They are highly complex systems that are intricately connected to various transportation modes, trade routes, and distribution channels.

They act as hubs that enable the smooth flow of goods between countries and continents. To create a more efficient and cost-effective trading route, logistics platforms have been established at strategic locations around the world to connect various trade routes. Moreover, logistics platforms are not just limited to transportation. They also incorporate several other aspects of trade, such as customs clearance, warehousing, and security.

These platforms have greatly simplified the process of international trade by providing a one-stop solution for all logistical needs. With advancements in technology, logistics platforms have become highly automated, reducing the time and effort required for trade operations. In summary, logistics platforms play a crucial role in facilitating global trade.

They are an integral part of the international network of trade routes, transportation modes, and distribution channels. These platforms have greatly simplified the trade process and become essential for the growth and development of economies around the world. With continuous technological advancements, logistics platforms will continue to evolve and play an even more significant role in the future of global trade.

Logistics platforms act as a bridge between the producers and the consumers, ensuring efficient and reliable transportation and logistics services. These platforms create an interconnected supply chain by providing a seamless transportation network, which is essential for any industry's growth. These result in faster delivery of goods, reduced costs, and ultimately increased revenue. Furthermore, logistics platforms also help businesses gain a competitive edge in the market. With the ever-increasing global competition, it has become crucial for companies to have a robust and efficient supply chain. This is where logistics platforms step in, providing businesses with access to the latest technology, advanced tracking systems, and real-time data. This allows companies to make informed decisions, reduce lead times, and improve overall efficiency, making them more competitive in the market. In addition to improved efficiency and competitiveness, logistics platforms also bring other benefits to businesses and industries.

They provide a centralised platform for managing the entire supply chain, from sourcing raw materials to delivering the final product to the end consumer.

1. Shipbuilding and repair

Shipyards serve as specialized facilities for the construction, repair, and maintenance of ships.They are critical facilities for the maritime industry's growth and development. They provide vital services for the construction, repair, and maintenance of ships, making them a crucial part of the industry's sustainability. Without shipyards, the industry would struggle to keep up with the increasing demand for ships and the constant need for repairs and maintenance.

Moreover, shipyards play a critical role in ensuring the safety and reliability of vessels that operate in global waters. They have the necessary expertise and equipment to construct and repair ships to the highest standards, ensuring that they can withstand the harsh conditions of the ocean.

This is especially important for ships that transport goods and people, as any malfunction or failure could have disastrous consequences. In addition to safety and reliability, shipyards also contribute to the efficiency of the maritime industry.

By providing top-notch services for ship construction and maintenance, they help increase the efficiency of shipping operations, reducing costs and improving overall performance.

The shipbuilding process starts long before any physical work is done, as the design and engineering phase is where the magic truly happens.

At this stage, naval architects and marine engineers come together and work on creating detailed plans and specifications for the vessel. These plans are essential, as they will serve as the blueprint for the entire shipbuilding process.

During this phase, every single aspect of the ship is carefully considered and planned out. This includes defining the ship's dimensions, hull form, structural layout, propulsion system, and outfitting arrangements.

130

All of these elements are critical to the vessel's success, and they must be meticulously planned out to ensure a smooth and efficient building process. The design and engineering phase is a critical step in shipbuilding, as it sets the foundation for the rest of the project.

The shipbuilding process can proceed to the next phase after finalizing the plans and specifications. This typically involves procuring materials, constructing the vessel, and conducting quality control checks along the way.

However, without the design and engineering phases, none of this would be possible. It is the first and most crucial step in bringing a ship to life, and it requires a high level of expertise and attention to detail. The success of the entire project depends on the work done during this phase, making it a crucial and exciting part of the shipbuilding process.

The next step after finalizing the design is to procure the necessary materials for the construction process. As with any construction project, there are a variety of materials that are required for the vessel to be built. These materials can range from steel plates and aluminium sheets to other structural components that are essential for the vessel's integrity and safety.

In order for the construction process to go smoothly, it is crucial that these materials meet specific quality standards and regulatory requirements. This ensures that the vessel will be able to withstand the harsh conditions of the sea and maintain its structural integrity.

The safety of the crew and passengers is of utmost importance, so it is imperative that all materials used in the construction meet these requirements. Once the materials have been procured, they are carefully inspected and tested to ensure that they meet the necessary standards.

This is a vital step in the process, as any subpar materials could compromise the entire project. The construction team must also carefully track and manage these materials to ensure that they are used efficiently and effectively in the building process. With the right materials and careful attention to detail, the construction of the vessel can move forward and bring the design to life.

The hull fabrication process is a crucial stage in shipbuilding. The hull structure is formed by the assembly of steel plates or aluminium sheets. This process is done according to the approved design plans, which ensures the integrity and strength of the hull.

The structural components are carefully cut and shaped, and they are then welded or riveted together to form the hull. This ensures a sturdy and reliable structure that can withstand the harsh conditions of the sea.

Advanced techniques, such as laser cutting and robotic welding, have revolutionised the hull fabrication process. These techniques have greatly improved efficiency and precision, resulting in higher-quality hulls. Laser cutting allows for more accurate and faster cutting of the steel plates, while robotic welding ensures consistent and strong welds.

This not only saves time but also reduces the margin of error, resulting in a more reliable and durable hull. The hull fabrication stage is a critical step in shipbuilding that requires skilled workers and advanced techniques. The hull's quality has a significant impact on the ship's overall performance and safety.

Therefore, it is essential to ensure that this stage is done with precision and accuracy to create a strong and reliable hull that can withstand the challenges of the open sea. Advanced techniques have greatly improved this process, making it more efficient and precise, ultimately resulting in safer and more durable ships.

Once the hull structure is complete, outfitting and installation work begins to equip the vessel with various systems, equipment, and components. This stage of the process is crucial, as it involves the installation of key elements that are necessary for the vessel to function properly.

The installation process is a complex and time-consuming one, as each piece of equipment must be carefully placed and integrated into the vessel's structure. The installation process begins with the installation of propulsion machinery, which is responsible for powering the vessel and allowing it to move through the water. This machinery is typically quite large and must be carefully placed and secured to ensure that it functions properly.

Once the propulsion machinery is installed, the vessel's electrical systems are next to be installed. These systems are responsible for providing power to all of the vessel's electrical components, such as lighting, navigation equipment, and communication systems. As the installation process continues, the vessel begins to take shape and become more functional. Piping is installed to transport various fluids throughout the vessel, such as fuel, water, and sewage.

The HVAC system is also installed, which is crucial for keeping the vessel's interior at a comfortable temperature. Accommodation facilities, such as cabins and living quarters, are also installed during this stage, providing the crew with a place to rest and relax.

Finally, navigation equipment, communication systems, and safety features are also installed, ensuring that the vessel is equipped to safely navigate the waters and communicate with other vessels.

The final assembly stage of a seaworthy vessel is a complex process that brings together the hull and outfitting components. All systems and components are integrated to create a complete and functional vessel. This stage requires the coordination of several trades, such as welders, electricians, plumbers, carpenters, and technicians. Each trade must work together to ensure that all systems are installed and tested according to the specifications.

The welders are responsible for joining the hull components together to create a sturdy and watertight structure. The electricians then install all electrical systems, including lighting, navigation equipment, and communication devices. Plumbers are tasked with installing the plumbing and sewage systems, while carpenters are responsible for building the interior of the vessel, including cabins, galleys, and storage spaces. Technicians, on the other hand, ensure that all systems are functioning properly and troubleshoot any issues that may arise during the testing phase.

The final assembly stage is critical in vessel construction because it brings together all of the different components and systems to create a seaworthy vessel.

It is a collaborative effort that requires the expertise of various trades to ensure that the vessel meets all specifications and is ready for the open sea. This stage's successful completion is a testament to the skill and dedication of the workers involved in the construction process.

Once the construction is complete, the ship is ready for the next phase of its journey. It will undergo a series of rigorous tests and trials to ensure its readiness for the open sea. This process is crucial, as it verifies the performance, safety, and compliance of the vessel with regulatory standards. T

These tests include a series of sea trials, where the ship's manoeuvrability, stability, speed, and sea keeping characteristics are evaluated under real-world conditions. This is a crucial step in the shipbuilding process, as it provides a true test of the ship's capabilities in the open ocean. In addition to the sea trials, functional tests are also conducted to ensure that all systems and equipment onboard are operating properly and efficiently.

Once all tests and trials have been completed, the ship is deemed seaworthy and ready for its journey. This marks the end of the construction process and the beginning of the ship's life at sea.

Shipyard workers play an essential role in the shipping industry. They help build new vessels, but they also maintain and repair existing ones.

Shipyards also perform repair, maintenance, and refitting work on existing vessels. This means that shipyard workers are responsible for routine maintenance tasks such as hull cleaning, painting, and equipment servicing.

These tasks are crucial to keeping vessels in top condition and preventing any potential issues while at sea. Furthermore, shipyard workers also handle major repair and overhaul projects to address structural damage, machinery failures, or system malfunctions. These repairs are crucial to maintaining the safety and reliability of ships.

Dry docks and floating docks are used to dock vessels for inspection, repair, and maintenance activities. These docks provide workers with a safe and stable environment for accessing and working on the ship's hull and other areas.

This is an essential aspect of shipyard work, as it ensures that ships are thoroughly inspected and repaired before being sent back out to sea. Overall, shipyards play a vital role in keeping the shipping industry running smoothly and safely.

Shipyards are essential facilities for ship construction and repair. They are equipped with a wide range of tools and equipment to support their activities. These facilities include workshops where skilled workers can assemble and construct ships. In addition to this, there are warehouses for the storage of materials and spare parts, cranes for lifting heavy objects, and slipways for launching ships into the water.

Furthermore, shipyards frequently have specialised equipment to carry out specific tasks. For example, welding machines are used to join metal parts together, while cutting tools are used to shape and cut materials. Shipyards also have lifting equipment to move heavy objects and testing facilities to ensure that ships are seaworthy. With these facilities, the construction and repair of ships can be carried out efficiently and effectively.

In addition, shipyards are equipped with berths and mooring facilities. These are used to secure ships while they are being worked on or waiting to be repaired. These facilities are essential for the safety of the ships and the workers, as they prevent accidents and damage to the vessels. With these facilities, shipyards can handle multiple ships at once, increasing their productivity and efficiency.

Shipyards are essential to the maritime industry, employing a diverse team of highly skilled professionals. Naval architects, who design and oversee ship construction, and marine engineers, who build ships to withstand the harsh conditions of the sea, comprise this team.

In addition, welders play a crucial role in the construction process by joining pieces of metal together to form the ship's structure, while fitters ensure that all components fit together seamlessly.

Electricians are responsible for installing and maintaining the ship's complex electrical systems, while painters give the vessel its final coat of paint. Riggers are also an important part of the team, responsible for moving heavy equipment and materials around the shipyard.

These skilled individuals' work extends to not only building ships, but also repairing and maintaining them. This involves following industry standards and regulatory requirements to ensure the safety and effectiveness of the ship. Customer specifications also play a crucial role, as each ship is uniquely designed to meet the specific needs of its owner.

This requires close collaboration between the different trade's people to ensure that the end product meets the customer's expectations. Their hard work and attention to detail ensure that ships are built to the highest standards, providing safe and reliable transportation for goods and people across the world's oceans. As the industry continues to grow, the demand for skilled workers in shipyards will only increase, making this a vital and rewarding career path for many individuals.

Shipyards are responsible for constructing, repairing, and maintaining ships, and they must comply with various regulatory requirements and industry standards.

These are in place to ensure ship quality, safety, and seaworthiness. To comply, shipyards must follow classification society rules, international conventions, flag state regulations, and local environmental laws. Failure to meet these standards could result in serious consequences, including fines and disruptions to operations.

Classification societies are organisations that establish and enforce technical standards for ship design, construction, and maintenance. They conduct surveys and inspections to ensure that ships are built and maintained to the highest safety and quality standards. International conventions, such as the International Convention for the Safety of Life at Sea (SOLAS), also play a crucial role in regulating ship construction and maintenance. These conventions set minimum safety and environmental standards that all ships must comply with when operating in international waters. In addition to these global regulations, shipyards must also adhere to flag state regulations. This refers to the laws and regulations set by the country where the ship is registered. These regulations cover everything from vessel design and construction to crew training and safety procedures.

Finally, shipyards must also comply with local environmental laws to ensure that their activities do not harm the surrounding ecosystem. These regulations can vary depending on the location of the shipyard and the type of work being done. Overall, compliance with these various requirements is crucial to maintaining the safety and integrity of the ships and protecting the marine environment.

1. Fishing, aquaculture and the fish industry
1. - Commercial fishing

Commercial fisheries are a critical component of our global economy, providing food and jobs for people all over the world. These large-scale operations are essential for harvesting fish and other marine resources for profit. Commercial fishing vessels use a variety of fishing equipment and techniques to catch and collect large quantities of fish for sale to markets, processing plants, and seafood distributors.

This industry plays a crucial role in meeting the ever-increasing demand for seafood as well as supporting the livelihoods of millions of people. The most common target species for commercial fisheries are high-value fish such as tuna, cod, salmon, shrimp, and shellfish. These species are highly sought after and command high prices in the market, making them lucrative targets for commercial fishermen.

However, this also means that they are at risk of overfishing, which can have significant impacts on the ocean's ecosystem and the livelihoods of those who depend on it.

As such, sustainable fishing practises and regulations are crucial to ensuring the long-term viability of these fisheries. Commercial fishing vessels use a variety of methods to catch fish, each with its own advantages and disadvantages. Trawling, for example, entails dragging a large net behind the vessel to catch fish, but it can also result in the capture of non-target species and damage to the seafloor. Long lining, on the other hand, uses a long line with baited hooks to catch fish, but it can also accidentally catch seabirds and other marine animals. Understanding these different methods and their potential impacts is essential for managing and regulating commercial fisheries effectively.

138

1. - Artisanal fishing

Artisanal fisheries, also known as small-scale or traditional fisheries, are a vital component of many coastal communities. Coastal communities and inland waterways often host these fisheries, where fishermen rely on traditional knowledge, hand-operated gear, and sustainable fishing practices to harvest fish in relatively small quantities.

Small boats and simple gear allow local fishermen to catch fish for subsistence or local markets. The small-scale operations allow the fishermen to manage their resources, conserve the environment, and support their livelihoods and food security.

The use of traditional knowledge and sustainable fishing practises also plays a significant role in preserving cultural heritage in many coastal regions around the world. Generations pass down these practices, ensuring the continuation of fishing traditions and techniques.

As a result, artisanal fisheries not only provide economic benefits but also hold cultural and social value within these communities.

They also contribute to the local food supply, providing a source of fresh and healthy seafood to the surrounding areas. Despite their importance, artisanal fisheries face many challenges, such as overfishing, pollution, and climate change.

These issues threaten the sustainability of these fisheries and the livelihoods of those who depend on them. Therefore, it is crucial to support and promote sustainable fishing practises in artisanal fisheries to ensure their continued existence and contribution to coastal communities. By doing so, we can not only preserve cultural heritage but also promote food security and sustainable resource management.

Angling is a popular pastime that involves catching fish for leisure or personal consumption. Many recreational fishermen use

139

fishing rods, reels, lines, and lures to catch fish in freshwater lakes, rivers, streams, and saltwater coastal areas. They may have different techniques and preferences when it comes to catching their target species, whether it is trout, bass, salmon, or even marlin. But for most anglers, the thrill of the catch and the time spent in nature are the main reasons for participating in recreational fishing.

There are different types of recreational fishing, including fly fishing, shore fishing, and boat fishing. Each type offers a unique experience and requires different skills and equipment. For example, fly fishing involves using a special rod and line to cast a lightweight fly or lure to attract fish, while shore fishing involves casting from the shoreline using a variety of baits and lures.

Boat fishing, on the other hand, allows anglers to travel to different spots and use different techniques to catch fish. Recreational fishing not only provides a fun and relaxing activity but also has important economic and social impacts. It supports local businesses, creates jobs, and contributes to the economy through equipment sales, tourism, and fishing licences.

Additionally, it brings people together and promotes a sense of community, as anglers often share tips and stories with each other while enjoying their time on the water. Overall, recreational fishing is a beloved pastime that offers many benefits and continues to be a popular activity worldwide.

1. - Aquaculture

Aquaculture, also referred to as fish farming, has gained popularity as a means of providing a sustainable source of seafood to meet the growing global demand for fish, shellfish, and aquatic plants. It involves the cultivation of various species of fish and other marine organisms in controlled environments such as ponds, tanks, cages, or raceways.

This method of fish production has become increasingly important in relieving pressure on wild fish stocks, which are depleting at an alarming rate due to overfishing. With the advancement of technology, aquaculture fisheries have expanded their range of farmed species to include not only traditional fish like salmon, tilapia, and catfish but also shellfish such as shrimp, oysters, mussels, and even seaweed.

These aquatic plants are not only a sustainable source of food, but they also have numerous environmental benefits, such as improving water quality and providing habitat for other marine species.

Additionally, aquaculture plays a vital role in global food production, providing a significant portion of the world's seafood supply. Despite its many advantages, aquaculture also faces its fair share of challenges, such as disease outbreaks and environmental impacts.

However, with proper management and regulations, these issues can be mitigated, and aquaculture can continue to be a crucial contributor to the world's food supply. As the demand for seafood continues to rise, aquaculture will play an increasingly vital role in meeting it sustainably.

Environmental certification schemes are becoming increasingly popular in the aquaculture industry. These voluntary programmes allow producers to prove their dedication to maintaining sustainable

141

and environmentally responsible practises. To achieve certification, producers must undergo rigorous third-party assessments and audits to ensure they are complying with specific environmental standards. These standards typically include measures to minimise pollution, conserve water resources, protect biodiversity, and prohibit the use of harmful chemicals or antibiotics. Two of the most well-known environmental certification schemes in aquaculture are the Aquaculture Stewardship Council (ASC) and the Global Aquaculture Alliance's Best Aquaculture Practices (BAP) certification. These programmes provide a way for consumers to easily identify and support environmentally responsible aquaculture operations. They also give producers a way to differentiate themselves in the market and showcase their commitment to sustainability. Environmental certification schemes not only benefit the environment and consumers, but they also benefit the aquaculture industry as a whole. By promoting sustainable practises and responsible resource usage, these programmes help ensure the long-term viability of the industry.

Some countries have recognised the importance of regulating and monitoring the safety and quality of aquaculture products. As a result, they have implemented mandatory certification processes or regulatory frameworks to ensure that their products are safe for consumption and adhere to environmental sustainability standards. These state certification typically cover a wide range of requirements, including food safety, hygiene, traceability, and labelling. In addition to meeting these standards, state-certified aquaculture products are also subject to regular testing, monitoring, and compliance checks by government agencies. This ensures that the products continue to meet the established standards and regulations, providing consumers with peace of mind that they are purchasing safe and high-quality products.

Furthermore, these certification processes also promote transparency and accountability within the aquaculture industry, as companies must comply with strict guidelines and regulations in order to obtain and maintain their certification.

Overall, this state certification plays a crucial role in ensuring the safety and quality of aquaculture products. They not only provide consumers with confidence in the products they are purchasing but also promote responsible and sustainable practises within the industry. These programmes, with continued monitoring and enforcement, can greatly contribute to the growth and success of the aquaculture industry, benefiting both producers and consumers.

Implementing sustainable management practises is key to reducing the environmental impact of aquaculture. By doing so, we can ensure that seafood production does not degrade our natural resources and ecosystems. This can be achieved by using environmentally friendly farming methods, such as integrated multitrophic aquaculture, which involves the cultivation of multiple species in a single system. This method not only minimises waste and pollution but also promotes biodiversity and enhances ecosystem resilience.

Aquaculture operations are a key component of the global food system that provide a reliable source of protein to a growing population. However, the expansion of aquaculture practises also comes with significant environmental concerns, most notably the discharge of effluents. These effluents, composed of excess feed, faeces, and chemicals, can have detrimental impacts on aquatic ecosystems. This is due to the fact that these effluents contribute to eutrophication, which is the excessive enrichment of nutrients in aquatic environments.

The effects of eutrophication can be far-reaching and severe. Not only does it lead to algal blooms that can choke off oxygen and suffocate fish and other aquatic organisms, but it also degrades the overall water quality. This, in turn, can have negative impacts on the health of aquatic biodiversity, disrupting the delicate balance of the ecosystem.

The growing demand for animal feed products, such as fishmeal and fish oil, for use in aquaculture diets has raised concerns about the negative impacts on marine ecosystems. The production of fishmeal and fish oil necessitates the use of wild-caught fish, which can contribute to overfishing and fish stock depletion. This unsustainable practice can also have detrimental effects on food webs and cause disruptions in marine ecosystems.

Furthermore, the reliance on wild-caught fish as feed ingredients can lead to habitat degradation and ecosystem disruption, exacerbating the negative impacts on marine environments.

Researchers are exploring sustainable aquaculture practices and alternative feed ingredients to address these concerns. This includes the use of plant-based proteins and oils, as well as the development of new technologies to produce fishmeal and fish oil from by-products and waste. These efforts aim to reduce the reliance on wild-caught fish and promote the long-term sustainability of aquaculture production.

1. - Subsistence fishing

Subsistence fisheries are a vital part of many people's lives around the world. These fisheries are run by local communities and indigenous groups, who use traditional fishing methods to harvest fish and other aquatic resources.

These methods include netting, trapping, spearing, and gathering in rivers, lakes, coastal waters, and marine environments. For these communities, subsistence fisheries are not just a means of obtaining food but also a way of life. They provide essential nutrition and protein, as well as cultural sustenance.

These fisheries are especially important for people living in remote and marginalised areas, who rely on them for their survival. Subsistence fisheries, on the other hand, face many challenges in today's world. The increasing industrialisation and pollution of water bodies have resulted in a decline in fish populations, making it harder for subsistence fishermen to catch enough food. In addition, governments often prioritise commercial fishing over subsistence fishing, making it difficult for these communities to access the resources they need. Despite these challenges, subsistence fisheries continue to be an important part of many cultures and play a crucial role in sustaining the lives of millions of people worldwide.

145

1. - Deep-sea fishing

Deep-sea fishing is incredibly important to the world of marine resources, and they are the mainstay of many industries worldwide. These fisheries are located in the deep ocean, far away from the continental shelf, and typically operate at depths of over 200 metres. This means that fishing vessels need to be specially equipped to handle these deep waters and the unique challenges that come with them.

Many deep-sea fisheries rely on advanced technology and equipment to find and catch their desired species. One of the main reasons deep-sea fisheries are important is because they target species that live in the ocean's deeper layers.

These species, such as various types of fish, crustaceans, and molluscs, are not typically found near the surface and require specialised fishing methods and equipment to be caught: Deep-sea fisheries play a crucial role in the seafood industry, offering a diverse range of species unavailable in shallower waters.

This allows for a diverse and sustainable supply of seafood for consumers around the world.

However, deep-sea fisheries also face their own set of challenges and concerns. Due to their remote location and depth, these fisheries are often difficult to monitor and regulate, leading to concerns about overfishing and potential damage to deep-sea ecosystems. As a result, it is critical that these fisheries are managed responsibly and sustainably to ensure the long-term health and viability of these valuable marine resources.

Deep-sea fisheries are an essential aspect of our global fishing industry. They are located in the vast, deep oceans, thousands of metres below the surface. The fishing grounds in these regions are most commonly found in underwater canyons, abyssal plains, and seamounts. However, due to the increased demand for fish

146

consumption, these fisheries are expanding and can now be found in other submarine features as well.

These fishing grounds are located far offshore, so special vessels and equipment are used to reach them. These vessels equip themselves with cutting-edge technology and equipment to navigate these fishing grounds.

However, it is not an easy task to fish in such deep waters.

Fishermen face many challenges, such as rough weather conditions, long hours, and the risk of encountering dangerous marine animals. Despite these challenges, deep-sea fisheries are a crucial source of seafood for the world, and they play a vital role in providing food security for many countries. As the world's population continues to grow, the demand for fish and seafood is also increasing.

Deep-sea fisheries are now facing the challenge of sustainability, as overfishing and destructive fishing practises can harm the delicate marine ecosystem. Therefore, it is essential to regulate these fisheries and promote sustainable fishing practises to ensure the long-term viability of these oceanic regions. The deep-sea fisheries provide a significant contribution to our economy and food supply, and it is our responsibility to protect and preserve them for future generations.

Deep-sea fisheries target a variety of species that are well adapted to the harsh and extreme conditions present in the deep ocean. This is due to the fact that deep-water fish such as orange roughy, grenadiers, black scabbard fish, and deep-sea cod have evolved over time to survive in the deep ocean's cold, dark, and high-pressure environment.

These species have unique and fascinating biological adaptations that allow them to thrive where other marine life would struggle to survive.

For example, deep-water fish have large eyes that are highly sensitive to light, which is important in the dark ocean depths. They also have specialised organs that help them detect and catch their prey.

In addition to these deep-water fish, other species that are commonly targeted by deep-sea fisheries include deep-sea shrimp, squid, crabs, and deep-water corals. These species are also well adapted to the challenging conditions of the deep ocean and play important roles in the marine ecosystem.

For instance, deep-sea shrimp are an important food source for many other species, and deep-water corals provide habitats for a diverse range of marine life. However, the continued exploitation of these species by deep-sea fisheries can have significant impacts on the health and functioning of the deep ocean ecosystem.

It is important for fishery management to carefully consider the impacts of deep-sea fishing and implement sustainable practises to ensure the long-term viability of these species and the ecosystems they support. Deep-sea fishing has been a source of controversy in recent years due to concerns over their sustainability and potential impacts on the marine environment.

While they provide a significant source of income and food for many coastal communities, their practises can have negative consequences if not managed properly. It is important for fisheries management to carefully monitor and regulate deep-sea fisheries to ensure that they are not causing irreversible damage to the deep ocean ecosystem. This includes setting catch limits, using sustainable fishing practises, and conducting thorough environmental impact assessments.

Deep-sea fisheries use a diverse set of fishing gear and techniques in order to capture the wide variety of fish and other organisms that can be found living at extreme depths. This includes methods such as bottom trawling, where a large net is dragged along the ocean floor to catch bottom-dwelling species, and long lining, where a line with numerous baited hooks is stretched across the seafloor.

One of the most common types of deep-sea fishing gear is gillnetting, where a large net with small mesh holes is anchored to the seafloor and catches fish as they swim into it.

Another popular method is pot fishing, where baited traps are set on the seafloor and then retrieved after a certain period of time. People often use these methods to target specific species or to minimize damage to the seafloor.

To increase clarity, use active voice: Deep-sea fishing has seen the development of more innovative and environmentally friendly methods in recent years. These methods include deep-sea fishing, which uses long lines to target specific species without damaging the seafloor, and pelagic trawling, which involves towing nets behind a vessel in the water column to catch fish near the surface.

These modern techniques not only reduce the environmental impact of deep-sea fishing but also result in higher-quality and more sustainable catches.

Deep-sea fisheries are an integral part of the fishing industry. It involves fishing in deep waters, which can be extremely challenging and risky due to various factors. One of the major challenges faced by deep-sea fisheries is the harsh weather conditions.

The unpredictable nature of the ocean can make it difficult for fishermen to navigate and operate their vessels. This can lead to dangerous situations and accidents, making deep-sea fishing a perilous profession. The long distances from port are another issue faced by deep-sea fisheries.

Unlike traditional fishing, in which the vessels stay near the coast, deep-sea fishing necessitates that fishermen travel far out into the ocean. This means longer and more exhausting trips, which can take a toll on both the fishermen and their equipment. Furthermore, the limited access to fishing grounds adds to the difficulty of deep-sea fishing.

With the increase in demand for seafood, fishermen are forced to go further and deeper into the ocean to find fish, making their job even more challenging. Apart from the physical challenges, deep-sea fisheries also have to consider the environmental impact of their activities.

Fishing in deep waters can have a significant impact on fragile deep-sea ecosystems. Certain fishing methods have the potential to damage or destroy cold-water corals, which are critical for the survival of many marine species. Furthermore, deep-sea fisheries may also harm vulnerable species with slow growth rates and low reproductive rates. This can have a detrimental effect on the delicate balance of the ocean's ecosystem.

Overfishing and its impacts on marine ecosystems have become a growing concern in recent years, leading to the implementation of strict regulations and management measures for deep-sea fishing. These regulations aim to ensure the sustainability of fish stocks and protect vulnerable marine ecosystems from further damage. As a result, international organisations such as the Food and Agriculture Organisation (FAO) and regional fisheries management organisations (RFMOs) have taken on a crucial role in establishing rules and guidelines for deep-sea fishing activities.

These regulations and measures are necessary to prevent the depletion of fish populations and maintain a balance in the ocean's delicate ecosystem. With technological advancements, deep-sea fishing has become more accessible and efficient, leading to an increase in fishing efforts. This has resulted in a decline in fish stocks, threatening the livelihoods of coastal communities and the food security of many nations. Therefore, the involvement of international organisations in regulating deep-sea fisheries has become crucial to ensuring sustainable fishing practises.

Despite these strict regulations, there are still concerns about the effectiveness of their implementation and enforcement. This is where the role of regional fisheries management organisations comes into play These organizations collaborate closely with coastal states and fishing nations to devise and execute measures that cater to the unique requirements of each region.

The researchers are constantly working to gain a deeper understanding of the intricacies of deep-sea ecosystems. Through their efforts, they hope to uncover more information about the complex web of life that exists in the ocean's depths.

This knowledge can then be used to assess the potential ecological impacts of deep-sea fishing and develop strategies to mitigate any negative effects. In addition to research, there are also ongoing conservation initiatives that have been put in place to protect the vulnerable marine habitats and species found in the deep sea. These efforts include the establishment of marine protected areas, which serve as safe havens for various marine creatures.

Along with this, there is also a push to promote sustainable fishing practises, which will help ensure that deep-sea fisheries can continue to thrive without causing harm to the surrounding environment. These conservation efforts are crucial to maintaining the delicate balance of life in the deep sea. Overall, the combined efforts of research and conservation are essential to preserving the health and well-being of deep-sea ecosystems.

By gaining a better understanding of these complex and mysterious habitats and implementing effective conservation measures, we can work toward a sustainable future for both the ocean and its inhabitants. It is important to continue these efforts and stay on track toward protecting and preserving these vital ecosystems.

1. - Illegal fishing

: Illegal, Unreported, and Unregulated (IUU) Fisheries: IUU fisheries are fishing activities that don't follow national or international rules. They usually use illegal, unreported, or unregulated fishing methods. These activities are a major threat to the sustainability of fisheries and the health of marine ecosystems. IUU fishing is a major global issue that affects both developing and developed countries, with an estimated global value of up to $23 billion per year.

The impact of IUU fishing extends beyond fish stock depletion. It also contributes to overfishing, habitat destruction, and the loss of biodiversity in our oceans. This has a ripple effect on the entire marine ecosystem, affecting not only the fish populations but also other marine species and their habitats. Furthermore, IUU fishing has negative economic and social impacts, including income loss for legitimate fishermen, displacement of local communities, and even human rights abuses on board fishing vessels.

Efforts to combat IUU fishing include enhanced monitoring and surveillance, improved enforcement, and international cooperation to promote responsible fishing practises and conservation measures. This requires a multi-faceted approach, involving governments, fishing industry stakeholders, and international organisations.

152

7 – Energy

7.1. - Wave energy

Ocean waves are generated by the wind blowing over the water's surface, so the kinetic energy they contain can be harnessed and converted into a usable energy source. These devices, known as wave energy converters (WECs), work by capturing the up-and-down motion of waves and transforming it into mechanical or electrical energy using a variety of different mechanisms. Some examples of these mechanisms include oscillating water columns, point absorbers, attenuators, and oscillating wave surge converters (OWSCs).

One common type of WEC is the oscillating water column, which uses a partially submerged chamber to capture the energy of passing waves. As the waves enter the chamber, the water level rises and falls, causing air to be forced in and out of a turbine. This movement of air drives the turbine, which in turn generates electricity. Point absorbers, on the other hand, consist of floating devices that are anchored to the seafloor and move up and down with the motion of the waves. This movement is then used to drive a generator and produce electricity.

Attenuators, which are typically long, slender structures that float on the water's surface, use the motion of the waves to drive hydraulic pumps that generate electricity. Lastly, OWSCs work by using the movement of waves to drive a piston or pump, which in turn drives a generator to produce electricity. To increase clarity, use active voice: These various mechanisms effectively capture and utilize wave energy as a renewable energy source.

153

7.2. - Tidal energy

The Earth's oceans are constantly moving, not just because of the wind and waves, but also because of the gravitational pull of the moon and sun. This movement creates a tremendous amount of kinetic energy in the form of tides.

Tidal energy converters (TECs) are designed to harness this energy and convert it into electricity. These devices are typically placed in areas where there is a significant difference between high and low tide, such as coastal areas, estuaries, and straits. One type of TEC is the tidal barrage system, which uses a dam-like structure to capture the energy from the tides.

As the tide rises, water is allowed to flow into a reservoir through turbines, generating electricity. When the tide goes out, the water is released back into the ocean, creating a continuous cycle of energy production. Another type of TEC is the tidal stream turbine, which looks similar to a wind turbine but is placed underwater to capture the energy from the moving tides.

Tidal kite systems, which use a large kite-like structure to capture the energy of the tides as they move back and forth, are also being developed. Tidal energy is a promising source of renewable energy, as it is predictable and reliable.

However, the technology is still in its early stages, and more research and development are needed to make it cost-effective and efficient. In addition, careful consideration must be given to the potential environmental impacts of installing TECs, as they can disrupt marine life and habitats. Despite these challenges, tidal energy has the potential to play a significant role in our transition to a more sustainable and clean energy future.

154

7.3 - Energy from currents

Ocean current energy is a renewable energy source that uses the power of ocean currents to generate electricity. This energy is harnessed by deploying underwater turbines and rotor systems that capture the kinetic energy of ocean currents, such as the Gulf Stream, the Kuroshio Current, and the Antarctic Circumpolar Current.

These powerful currents are constantly moving due to the Earth's rotation and differences in water temperature and salinity, making them a reliable source of energy.

The technology for extracting energy from ocean currents is still relatively new and is constantly being improved. Innovations such as flexible turbine blades and advanced tracking systems have increased the efficiency and reliability of ocean current energy systems. This form of renewable energy has the potential to provide a significant amount of electricity to coastal communities, reducing their reliance on fossil fuels and helping to combat climate change.

When harnessing ocean currents for energy, there are potential environmental impacts to consider, as with any form of energy production. Careful planning and monitoring are necessary to minimise these impacts and ensure the sustainability of this renewable energy source. With continued research and development, ocean current energy has the potential to play a significant role in our transition to a cleaner and more sustainable future.

155

7.4 - Ocean thermal energy

The concept of OTEC, or Ocean Thermal Energy Conversion, is an innovative way to generate electricity using the temperature difference between warm surface waters and cold deep waters in tropical regions. This new technology carries the promise of sustainable and renewable energy. OTEC systems typically consist of a surface heat exchanger, a working fluid, a turbine-generator unit, and a cold water pipe extending into the deep ocean. This system works by utilising the warm surface water to vaporise the working fluid, which then expands and drives a turbine to produce electricity. The design of OTEC systems maximizes the temperature difference between warm and cold waters.The greater the temperature difference, the more efficient the system becomes. Furthermore, OTEC systems have a minimal impact on the environment, making them a more viable and sustainable option for energy production. The surface heat exchanger is responsible for absorbing the thermal energy from the warm surface water, while the cold water pipe is used to transport the cold, deep water to the surface.

The working fluid, such as ammonia or propane, is a crucial component in this process, as it is responsible for driving the turbine and producing electricity. With the increasing demand for energy and the growing concern for the environment, OTEC has become a promising solution for both.

This technology, which utilises the vast resources of the ocean, has the potential to provide clean and renewable energy for years to come. The development and continued improvement of OTEC systems are critical to reducing our dependence on fossil fuels and mitigating the effects of climate change. With its many benefits and potential, OTEC is a technology that should be explored and invested in for a sustainable future.

156

7.5 - Salinity gradient energy

The process of producing salinity gradient energy, also referred to as osmotic power or blue energy, is primarily achieved by taking advantage of the varying levels of salt concentration between seawater and freshwater. This disparity is addressed by semi-permeable membranes that separate salt ions from water molecules.

In doing so, a pressure difference is created that can then be harnessed to drive turbines and ultimately generate electricity. Salinity-gradient power plants are designed to effectively maximize this energy source's potential.

These plants use semi-permeable membranes placed between two chambers, one containing seawater and the other containing freshwater. As the salt ions in seawater are drawn to the freshwater chamber, the pressure in the seawater chamber increases, generating a flow of water that can be used to drive turbines and convert the energy into electricity.

The concept of salinity gradient energy has received significant attention in recent years due to its potential to generate clean and renewable energy. With the world's population and energy demands continuously increasing, alternative energy sources such as salinity gradient energy are becoming more crucial in ensuring a sustainable future. As research and development in this field continue to progress, we can expect to see a significant increase in the utilisation of this innovative source of power.

157

7.6 - Offshore wind turbines

Offshore wind turbines are a source of renewable energy that are designed to harness the power of wind and convert it into electricity. Unlike their onshore counterparts, offshore turbines are specifically engineered to withstand the harsh and ever-changing marine environment.

These turbines are constructed to be durable, reliable, and long-lasting, and they are often significantly larger and more powerful than onshore turbines. The structure of offshore turbines is similar to that of their onshore counterparts. They consist of four main components: a tower, a nacelle, rotor blades, and a foundation. The tower is a tall structure designed to hold the rest of the turbine above the water. The nacelle is a housing unit that contains the generator and gearbox, which are responsible for converting the kinetic energy of the wind into electrical energy. The nacelle attaches the rotor blades, designed to capture and rotate the wind.

Finally, the foundation is a sturdy structure that anchors the turbine to the seabed and keeps it stable in the turbulent marine environment. Offshore wind turbines are an important source of clean and renewable energy, and their potential for growth and development is immense. They are constantly evolving and improving, with new technologies being developed to make them more efficient, reliable, and cost-effective.

Offshore wind farms' location is typically determined by a variety of factors. Each wind farm's location is unique and takes into consideration wind resources, water depth, seabed conditions, proximity to the shore, and regulatory requirements. All of these factors contribute to determining the most suitable location for the wind farm. Depending on these factors, offshore wind farms can be situated in different areas, such as coastal waters, shallow seas, or deeper offshore areas.

158

159

The location of the wind farm can also vary, with some being situated near the coastline, also known as near shore, while others are located further offshore, in mid-range or deepwater areas. The distance of these wind farms from the shore can vary greatly, ranging from a few kilometres to dozens or even hundreds of kilometres. Ultimately, the location of an offshore wind farm plays a crucial role in its success.

The factors considered in determining its location must be carefully evaluated to ensure the wind farm is able to operate efficiently and adhere to regulatory requirements. The varying locations of offshore wind farms also showcase the versatility and adaptability of this renewable energy source, making it a viable option for coastal communities looking to reduce their carbon footprint.

Offshore wind energy is a rapidly growing sector in the renewable energy industry; however, it still faces a multitude of challenges. One of the main challenges is the higher upfront costs compared to other energy sources. Moreover, the installation and maintenance of offshore wind farms is a complex and costly process. The logistics of offshore wind projects require specialised vessels and equipment, which further adds to the cost. As a result, many

investors are hesitant to allocate funds to offshore wind projects.

The harsh marine conditions are another challenge facing offshore wind energy. The turbines are situated in the open sea, making them susceptible to strong winds, rough seas, and extreme weather events. This can result in damage to the turbines, leading to costly repairs and maintenance.

Furthermore, the seabed constraints in certain areas make it difficult to install the turbines securely. This not only adds to the cost but also poses a safety risk for the workers involved in the installation process.

The potential conflicts with marine ecosystems, shipping lanes, fishing activities, and other maritime users are also a major concern in the development of offshore wind energy. The construction and operation of offshore wind farms can disrupt the natural habitat of marine animals and impact their migration patterns. It can also interfere with shipping routes and fishing activities, causing conflicts with other users of the sea.

8. Minerals

The process of extracting and exploiting these resources is not without its challenges. Technical limitations, environmental concerns, and regulatory complexities all play a role in shaping the potential of sea mineral resources.

One of the major challenges facing the industry. The depths at which these resources are found present a unique set of obstacles that must be overcome in order to extract them effectively. In addition, there are concerns about the environmental impacts of these activities on marine ecosystems and the delicate balance of biodiversity in the ocean.

These concerns must be carefully addressed in order to ensure sustainable resource management. Furthermore, the exploitation of sea mineral resources also brings up questions of ocean governance and international legal frameworks. The delicate balance between reaping the potential benefits of these resources and protecting the marine environment requires careful consideration of social, economic, and environmental factors.

This must be accompanied by stakeholder engagement, scientific research, and policy coordination at both the national and international levels. As a result, the search for sea mineral resources must be approached with caution and consideration in order to strike a balance between progress and preservation.

Marine environments are home to a vast range of sand and gravel deposits, which are formed through a variety of natural processes. One of the most common ways these deposits are created is through the constant erosion and sedimentation caused by the powerful forces of wind and water.

As rivers flow towards the ocean, they carry with them large amounts of sediment, which eventually settles onto the sea floor. This sediment is then transported by ocean currents and waves,

161

eventually forming deposits in various locations. Underwater sandbanks are one of the most common places to find marine sand and aggregates. These are large underwater formations that are created by the constant movement of water and the gradual accumulation of sediment.

Another common location is on beaches, where the constant crashing of waves against the shore helps break down larger rocks and create smaller particles of sand and gravel. The powerful forces of the ocean constantly move and shape nearshore sediments, located close to the shoreline, providing a common source of marine sand and aggregates.

Offshore shoals are another important location where sand and gravel deposits can be found. These are shallow oceanic areas where the sea floor rises, causing sediment to accumulate and form large deposits.

These deposits are often incredibly valuable, as they can be used in a variety of industries, such as construction.

The exploitation of marine sand and aggregates presents a number of complex challenges and considerations. While the potential benefits of these resources are vast, the negative impacts on the environment must be carefully evaluated and addressed.

This includes potential habitat disturbance, sedimentation, and changes to coastal dynamics, all of which can have a significant impact on marine ecosystems and coastal processes. Therefore, conducting thorough assessments and implementing effective management strategies is crucial to minimize any adverse effects.

In addition to environmental concerns, there are also important regulatory and social considerations that must be taken into account when exploiting marine sand and aggregates. There is a need for robust regulatory frameworks and permitting processes, which are essential for ensuring responsible and sustainable management of these resources.

8.1 - Polymetallic nodules

Polymetallic nodules are not only important due to their high metal content but also because they are an important source of rare earth elements (REEs).

These elements are used in a variety of modern technologies, such as smart phones, electric cars, and renewable energy sources. In fact, the demand for REEs is rapidly increasing, making the exploration and extraction of polymetallic nodules more relevant than ever. The formation of these nodules is a complex process that takes place over millions of years. They are formed when minerals and metals in seawater settle and accumulate around a small nucleus, such as a shark tooth or a piece of rock.

As the nodules grow, they trap other minerals and metals in their layers, resulting in their high concentrations of valuable elements. This slow formation process makes these nodules a non-renewable resource, highlighting the importance of responsible and sustainable extraction methods.

Despite their potential economic benefits, the exploration and extraction of polymetallic nodules also raise concerns regarding potential environmental impacts. The seafloor and sediment released during the extraction process can harm deep-sea ecosystems and disrupt important oceanic processes.

163

8.2 - Manganese nodules

Cobalt-rich ferromanganese crusts, also known as manganese nodules, occur as crust-like deposits on underwater mountains, seamounts, and rocky ridges in the deep ocean. These crusts are of great interest to the scientific community due to their significant concentrations of cobalt, manganese, nickel, and other metals, as well as rare earth elements (REEs).

Scientists are particularly interested in the potential economic value of these crusts, as cobalt and REEs are used in the production of high-tech devices such as smart phones and electric car batteries. The formation of these crusts is a slow and gradual process that takes place over millions of years. As seawater passes through the cracks and crevices on the seafloor, it carries with it dissolved minerals and metals. When the water reaches the cold and dark environments where these crusts form, it cools, and the minerals and metals precipitate out of the water and onto the seafloor.

This process is similar to the formation of stalactites and stalagmites in caves, but on a much larger scale and in a completely different environment. Despite their economic potential, harvesting these crusts is not without controversy. The process of mining these crusts involves scraping the seafloor, potentially damaging delicate ecosystems that have formed over millions of years.

There are also concerns about the impact on marine life, as many organisms rely on these crusts for food and habitat. As research on cobalt-rich ferromanganese crusts continues, scientists and policymakers must carefully consider the potential benefits and risks of mining these valuable resources from the deep ocean.

164

8.3 - Seabed massive sulphides

Seafloor massive sulphides (SMS) are of great interest to scientists and mining companies alike. These hydrothermal vent systems are discovered along mid-ocean ridges and volcanic seafloor areas. They are formed by the discharge of hot, mineral-rich fluids from beneath the Earth's crust.

Sulphuric minerals, including copper, zinc, lead, silver, gold, and others, are abundant in these fluids. SMS's discovery has opened up new possibilities for mining and extracting valuable mineral resources from the ocean floor. These high-grade ore deposits are highly sought-after by mining companies. However, extracting these valuable metals from the seafloor is a challenging task.

The harsh conditions of the deep sea, along with the complex geology of the seafloor, make mining for SMS a difficult and expensive process. Despite the challenges, the potential for valuable mineral resources from SMS has led to increased research and exploration efforts. Scientists are studying SMS formation and distribution to better understand how these deposits are created and where they are most likely to be found.

This knowledge will be crucial in developing sustainable ways to extract these resources without causing harm to the delicate ocean ecosystems. With ongoing research and advancements in technology, the future of SMS mining looks promising.

165

8.4 - Placer

Placer deposits are concentrations of heavy minerals and precious metals. They are called placer deposits because the precious metals are "placed" in one location, creating a highly concentrated source for minerals and metals. These minerals are often found in coastal and marine sediments, beaches, and near-shore environments. Placer deposits are highly sought-after by miners and companies alike. These deposits are formed by the erosion and transport of minerals from terrestrial sources to the seabed. This means that there is a high concentration of valuable minerals in one location, making it easier and more profitable to extract.

The process of forming placer deposits is complex and can take thousands of years. It begins with the erosion of minerals from rocks and soil on land. These minerals are then transported by rivers, currents, and wave action to the seabed.

Over time, these minerals accumulate and form highly concentrated deposits, which we know as placer deposits. This process also explains why these deposits are often found in coastal and marine environments, as they are formed by the movement of water.

166

8.5 - Rare Earth Elements

Rare earth elements (REEs) are becoming increasingly valuable and sought-after in today's ever-changing technological landscape. These 17 chemically similar elements possess unique properties that make them essential in high-tech industries ranging from electronics to renewable energy and defence technologies.

Their presence in these industries is undeniable, and the demand for these elements continues to grow. Various marine mineral deposits, such as polymetallic nodules, ferromanganese crusts, and seafloor massive sulfides, contain REEs.

These deposits are often located in remote and challenging environments, making their exploration and extraction a difficult and expensive task. However, the potential benefits of accessing these REEs make them a highly sought-after target for mineral exploration.

As technology advances and the demand for REEs grows, the race to discover and extract these elements from the ocean floor intensifies. The unique properties of REEs make them essential components in many advanced technologies. For instance, they are crucial in the production of electric and hybrid vehicles, wind turbines, and solar panels. Without these elements, the efficiency and performance of these technologies would be greatly diminished. Additionally, defence technologies like missiles, radars, and precision guided weapons use REEs. As our society becomes more reliant on technology, the demand for REEs will only continue to increase, making their exploration and extraction a crucial

undertaking.

167

8.6 - Energy minerals

The oceans are home to a vast array of natural resources, including many valuable energy minerals. Indeed, the potential for energy production is one of the most exciting aspects of the world's oceans. For example, methane hydrates, also known as gas hydrates, are a type of gas deposit found in the seabed.

These deposits are commonly found in marine sediments, subsea formations, and continental shelves. These energy resources have tremendous potential for commercial development and may be used to meet global energy demand. In addition to methane hydrates, there are many other energy minerals found in the sea. For instance, gas and oil deposits are also present in marine sediments and other underwater formations.

9 - Biotechnology

The use of marine organisms and materials has proven to be very beneficial in the development of products and solutions across different sectors, such as industry, healthcare, agriculture, and environmental conservation. This branch of science, known as biotechnology, is constantly evolving and has opened up many opportunities for research and development. With the increasing demand for sustainable practises and resources, marine biotechnology has emerged as a promising field that can provide innovative solutions.

One of the main advantages of marine biotechnology is its ability to harness marine organisms' unique characteristics and rich biodiversity: Algae, bacteria, fungi, invertebrates, and marine plants, among others, have adapted to survive in extreme environments and possess unique properties that can serve various purposes. For instance, scientists have discovered that certain types of algae can generate biofuels, and bacteria can aid in the development of antibiotics.

168

Marine biotechnology is a critical field that has gained increasing interest and attention in recent years. As the name suggests, it entails the use of marine organisms for the discovery, isolation, and characterisation of bioactive compounds and natural products.

These compounds have diverse applications in the pharmaceutical, nutraceutical, and cosmetic industries. With the increasing demand for alternative drug sources, marine biotechnology has emerged as a promising avenue for the development of novel therapeutic agents. Marine biotechnology's potential stems from the fact that marine organisms are known to produce a wide range of bioactive molecules. These molecules have been found to possess a variety of therapeutic properties, making them an attractive target for drug discovery.

Some of the most promising compounds obtained from marine organisms include antibiotics, anticancer agents, anti-inflammatory compounds, antioxidants, and neuroprotective substances. These bioactive molecules are not only effective in treating various diseases but also have fewer side effects compared to traditional drugs. Furthermore, their unique structures and properties make them an appealing option for developing new drugs.

Marine biotechnology has revolutionised the development of sustainable industrial practises and applications. The vast and diverse marine ecosystem provides a plethora of resources that can be harnessed to produce by-products, biomaterials, and biocatalysts.

People highly value these natural resources because they provide a more environmentally friendly and cost-effective substitute for traditional chemical-based products.

With the increasing demand for sustainable solutions, marine biotechnology has become a crucial player in the bioremediation, wastewater treatment, biofuel production, and bioprocessing industries.

In these industries, the use of marine-derived enzymes, proteins, polymers, and biopolymers has been a major breakthrough in biotechnology. These biomolecules have unique properties that make them highly efficient and versatile. For instance, marine enzymes are known for their ability to function in extreme conditions, such as high temperatures and salinity, making them ideal for industrial processes.

Conversely, the production of biodegradable plastics and wound dressings has utilized marine biopolymers like chitin and alginate. Overall, using biomolecules from the ocean has not only helped make products that last longer, but it has also made industrial processes less harmful to the environment.

Marine biotechnology has major implications for marine pollution, contamination, and environmental degradation. : This creative method uses many methods, like creating biosensors, biomarkers, and bioremediation technologies, to keep an eye on, evaluate, and fix the damage that people are doing to marine ecosystems and coastal habitats.

By harnessing the power of biotechnology, scientists are able to better understand and address the complex challenges facing our oceans. One particularly important aspect of marine biotechnology is the use of bioremediation techniques. This involves the use of microorganisms, plants, and other biological agents to remove pollutants from marine environments.

Bioremediation can effectively and efficiently clean up contaminated areas, reducing the harmful impacts of pollution on marine life and coastal communities. This approach has already been successfully applied in a number of cases, and ongoing research continues to improve and expand its capabilities. In addition to bioremediation, marine biotechnology also encompasses other important strategies, such as phytoremediation and biofiltration.

These methods involve the use of plants and other organisms to absorb and break down pollutants, providing a more sustainable and environmentally friendly approach to remediation. By utilising a combination of biological tools, marine scientists are able to develop comprehensive solutions to the complex challenges facing our oceans.

It entails the exploration and utilization of marine organisms' genetic resources, microbial diversity, and functional genomics for biotechnological applications. This means that scientists are studying the genetic makeup of different marine organisms to find new ways to use them in industries such as medicine, food, and energy production.

One of the main goals of marine biotechnology is to discover new genes, enzymes, and metabolic pathways from marine microbes and macroorganisms that could have biotechnological and industrial relevance. This process is known as bioprospecting, and it involves the careful study and analysis of different marine species to identify their unique genetic traits. These traits could then be harnessed for a variety of applications, such as creating new medicines or developing more efficient ways to produce renewable energy. Through bioprospecting, scientists hope to unlock the full potential of the vast and diverse world of marine organisms. Marine biotechnology has vast potential, and the possibilities for its applications are endless.

Has a significant impact on biomedical research, drug discovery, and healthcare innovation. These marine-derived products are proving to be extremely beneficial in treating a range of health challenges, including infectious diseases, cancer, neurodegenerative disorders, and tissue regeneration.

The development of marine biomaterials is one of the most promising areas in marine biotechnology. These materials are derived from marine organisms and have unique properties that make them ideal for use in medical procedures. Additionally, marine-derived drugs have shown great potential in treating cancer and other diseases, providing new options for patients who may have exhausted traditional treatments. Certain marine animals that regenerate damaged tissues have inspired the development of regenerative therapies that could potentially help patients with injuries or degenerative diseases. With continuous advancements in this field, the potential for marine biotechnology to revolutionise healthcare is immense.

We have put in place international agreements, national legislation, and regional initiatives to promote responsible bioprospecting, benefit-sharing, and sustainable utilization of marine biodiversity for biotechnological applications.

One of the key focuses of these regulatory frameworks is to ensure that both the marine environment and its inhabitants' rights are protected.

This encompasses ethical guidelines that delineate the ethical factors to be considered during research and the utilization of marine genetic resources.

Additionally, intellectual property rights play a significant role in the marine biotechnology sector, as they dictate who has the right to access and profit from these valuable resources.

10 - Seaweed

Seaweed has been a vital part of the marine ecosystem for centuries. It is a diverse group of marine plants that can be found in oceans, seas, and coastal waters all around the world. The term "seaweed" encompasses a wide range of species, from tiny phytoplankton to large, multicellular algae that can form extensive underwater forests or beds.

This variety of seaweed species plays a crucial role in maintaining the balance of marine ecosystems. One of the key ecological roles of seaweed is to provide habitat and food for a wide range of marine organisms. Many species of fish, crustaceans, and other marine animals rely on seaweed for shelter and as a food source. Additionally, seaweed contributes to biodiversity by creating a diverse and complex underwater environment, which in turn supports a wide range of marine life.

This makes seaweed a vital component of healthy and thriving marine ecosystems. Apart from its role in supporting marine life, seaweed also contributes to nutrient cycling and ecosystem stability.

As seaweed absorbs nutrients from the water and stores them in its tissues, it helps maintain the balance of nutrients in the surrounding water. This, in turn, supports the growth of other marine plants and animals.

Seaweed also helps stabilise the ocean floor by providing a natural barrier to strong ocean currents and waves, preventing erosion, and protecting coastal areas from storms.

Seaweed has been an important part of human life for centuries People have used seaweed in a variety of ways, from food to medicine.With a wide range of seaweed options such as nori, kelp, dulse, and wakame, it has been consumed as a nutritious food source in many cultures worldwide.

173

The benefits of seaweed extend beyond its consumption; it has also found its place in a variety of industrial applications. With a high nutritional value, seaweed has been a staple food for many communities over the years. It has been a reliable source of food during tough times, such as famines, and has played a crucial role in sustaining human life.

In addition to its consumption, seaweed serves as a natural fertilizer in agriculture, enhancing the quality of the soil.

This has helped increase crop yields and boost agricultural production. The usefulness of seaweed does not end here. It has also found its way into various industries, such as food, cosmetics, and pharmaceuticals. Seaweed extracts and derivatives have been used in food additives, giving them a natural touch and enhancing their nutritional value The cosmetic industry uses seaweed in skincare products due to its anti-inflammatory and anti-aging properties.

Its extracts have also been used in pharmaceuticals, showing promising results in treating various diseases.

Seaweed cultivation, also known as mariculture, has gained popularity as a sustainable form of commercial farming. This process involves the farming of specific seaweed species in controlled marine environments, such as coastal waters, bays, or offshore areas. It not only provides a sustainable source of biomass and bioactive compounds but also helps reduce the pressure on wild seaweed populations, promote coastal livelihoods, and foster economic development.

Seaweed farming has emerged as a lucrative industry in recent years, with a growing demand for seaweed-based products in various sectors, such as food, cosmetics, pharmaceuticals, and biofuel. The controlled environment in which seaweed is cultivated ensures high-quality and consistent production, making it an ideal choice for commercial purposes.

175

Seaweed farming is considered environmentally friendly because it helps to mitigate the negative impacts of overfishing and pollution on marine ecosystems. Seaweed farming is beneficial not only for the economy and the environment but also for the local communities that are involved in this practice.

By providing a sustainable source of income, it promotes the growth of coastal livelihoods and helps in the economic development of these communities. Furthermore, seaweed cultivation has other positive impacts, such as the absorption of excess nutrients from the water, which helps improve water quality and promote marine life growth.

Seaweed cultivation provides many environmental advantages, and one of them is carbon sequestration. The process entails absorbing carbon dioxide from seawater via photosynthesis. This contributes to ecosystem restoration and helps reduce ocean acidification and climate change. The seaweed absorbs the excess carbon dioxide from the seawater and converts it into oxygen, thus helping to offset the harmful effects of carbon emissions.

Furthermore, seaweed farming also plays a crucial role in improving water quality. The process of assimilating excess nutrients, such as nitrogen and phosphorus, from coastal waters helps reduce the risk of eutrophication and algal blooms. These nutrients are often found in excess due to agricultural runoff and sewage discharge. Seaweed acts as a natural filter, absorbing nutrients, preventing them from entering the ocean, and causing harm to marine life. This not only benefits the environment but also has a positive impact on the economy by preserving marine resources.

1. - Maritime Works

Maritime work is critical for ensuring navigation safety and protecting the coastal environment. These projects are crucial to ensuring the sustainability of marine resources and the well-being of coastal communities.

These include the construction of ports and harbours; the maintenance and renovation of existing infrastructure; and the development of new facilities to meet the growing demand for maritime transport.

Furthermore, maritime operations are critical for economic development because they provide access to new trade routes and promote the growth of international commerce. In addition, these projects are critical for protecting coastal areas from the impacts of climate change, such as sea level rise and extreme weather events.

They involve the implementation of measures to prevent coastal erosion and flooding, as well as the restoration of marine habitats and ecosystems. Overall, maritime work plays a vital role in the sustainable development of coastal and marine environments. They require a multidisciplinary approach and collaboration between different stakeholders, including engineers, environmental managers, and local communities.

176

1. - Ports and harbours

Ports and harbours are vital to the global economy and an essential component of maritime trade and transportation. These facilities are the central point of convergence between land and sea for all maritime activities. Its functions include the construction, expansion, and maintenance of various maritime works such as wharves, docks, piers, quays, breakwaters, jetties, and navigation channels.

These works are designed to accommodate ships of various sizes, cargo handling equipment, and other port facilities. The development of ports and harbours is an ongoing process that aims to improve vessel access, berthing capacity, cargo handling efficiency, and port security.

With the constant growth in global trade and shipping, these facilities must be continually expanded and upgraded to meet the industry's demands. However, this development must also take into account the environmental impacts of these projects.

We must implement measures to minimize these impacts and preserve the delicate balance of the marine ecosystem.

Furthermore, port development projects have significant economic benefits, not just for the local community but also on a global scale. These facilities create job opportunities, boost international trade, and contribute to the growth of the economy. However, it is crucial to ensure that these developments are sustainable and do not negatively affect the environment. With careful planning and execution, ports and harbours can continue to serve as essential hubs for maritime activities while promoting economic growth.

177

1. - Coastal protection

Coastal protection works are a critical element in the battle against nature's destructive forces. The coastal communities have been facing the brunt of coastal erosion, storm surges, and the looming threat of sea-level rise for decades. The consequences of such events are severe, ranging from damage to infrastructure to the destruction of natural habitats.

To combat these challenges, coastal defence structures, including seawalls, revetments, groynes, breakwaters, and beach nourishment projects, have been implemented. These structures play a significant role in stabilising shorelines, reducing wave energy, and preventing coastal flooding and erosion. However, these structural measures alone are not sufficient to protect the coast.

That is where integrated coastal management approaches come into play. Integrated coastal management approaches combine both structural and non-structural measures to enhance coastal resilience, preserve natural habitats, and maintain coastal processes.

These measures include the restoration of natural habitats, beach nourishment, and the creation of buffer zones. By implementing these measures, coastal communities can mitigate the impacts of coastal erosion and storm surges, providing a more sustainable and long-term solution.

178

1. - Dredging and marine excavation

Dredging and marine excavation are essential to the safety and efficiency of navigation channels, ports, and harbours. By removing sediment, sand, silt, and debris from water bodies, these activities ensure that water depths are maintained and that ships, vessels, and maritime traffic can access these areas safely.

This is crucial for the smooth operation of transportation and trade, as well as for the protection of marine ecosystems. There are various techniques that are used in dredging projects, each with their own unique purpose and method.

Hydraulic dredging, for example, involves using high-pressure water jets to break up and remove sediment from the bottoms of water bodies. Suction dredging, on the other hand, uses large pumps to suck up sediment and debris from the bottom of waterways.

Clamshell dredging and bucket dredging are also commonly used methods, which involve the use of large mechanical buckets to scoop and remove sediment from the bottom of water bodies.

These techniques necessitate skilled operators and specialized equipment, and they must be carefully planned and executed to ensure the project's safety and success. In addition to maintaining waterway access, dredging and marine excavation also play an important role in protecting the environment.

By removing excess sediment and debris, these activities improve water quality and help prevent the buildup of harmful pollutants. They also create and maintain habitats for marine life, which is essential for a healthy and diverse ecosystem. Proper planning and execution of dredging projects is crucial to ensuring that the delicate balance of the marine environment is maintained while also meeting the needs of human activities and industries.

179

1. - Offshore Structures

Offshore structures are essential for the extraction, production, and transport of offshore resources and energy. These structures are strategically designed, fabricated, installed, and maintained in offshore environments. They include a variety of facilities, such as offshore oil and gas platforms, subsea pipelines, underwater cables, and marine renewable energy infrastructure.

The construction of offshore structures is a complex process that requires advanced engineering techniques and specialised equipment. It involves all aspects of design, fabrication, installation, and maintenance.

The platforms, foundations, subsea structures, and support systems are engineered to withstand harsh weather conditions and strong ocean currents.

The offshore environment presents many challenges, and the structures must be designed to handle them all. Moreover, these offshore structures play a crucial role in the economy. They provide a stable energy supply and create job opportunities for many people.

The construction of such facilities is a significant investment and requires careful planning and execution. The engineers must take into account a variety of factors, such as environmental impact, safety, and cost-effectiveness. It is essential to maintain these structures regularly to ensure their integrity and safety. In conclusion, offshore structures are vital for the production and transportation of offshore resources and energy, and their construction is a complex process that requires expertise and precision.

Maritime work is critical for the success of marine renewable energy sources, including offshore wind, tide, wave, and ocean thermal energy. These renewable energy sources are crucial for

180

reducing our reliance on fossil fuels and mitigating the effects of climate change.

Without specialised marine infrastructure, such as support structures, mooring systems, and submarine cables, these energy sources would not be able to harness and deliver renewable energy from marine environments to onshore grids.

Offshore wind farms are one example of how maritime work is used to harness renewable energy. These farms are located in the ocean, where strong winds can be captured by wind turbines. However, specialized support structures and mooring systems must securely attach these turbines to the ocean floor for proper operation.

Additionally, subsea cables are used to bring the energy from the turbines to onshore grids, where it can be distributed to homes and businesses. Maritime activities not only play a crucial role in the development of renewable energy but also have a positive impact on the environment.

By utilising these energy sources, we can reduce our carbon footprint and decrease our reliance on non-renewable resources.

1. - Hydrographical surveys

Hydrographical surveys are also used to map and monitor changes in underwater terrain over time. This is important because underwater terrain can shift and change due to natural events, such as storms and erosion, as well as human activities, such as dredging and construction.

By conducting regular hydrographical surveys, scientists can identify and analyse these changes, providing valuable information for coastal and environmental management. In addition to mapping the depth and shape of the ocean floor, hydrographical surveys can also reveal important features of underwater terrain, such as seamounts, canyons, and trenches. These features are not only visually stunning, but they also provide critical habitats for a wide variety of marine life.

By accurately mapping and understanding these features, scientists can better protect and manage these unique and important ecosystems.

Hydrographical surveys are also important in the exploration and exploitation of natural resources. By mapping the seabed, scientists can identify potential areas for oil and gas extraction, mineral mining, and renewable energy development. This information is vital for sustainable resource management and can also help prevent conflicts over resource usage in shared water bodies.

Topo-hydrographical surveys are essential for understanding the world around us. These surveys combine two different surveying methods, hydrographical and topographic, to provide detailed data on both underwater and coastal environments. They are able to do this by using specialised surveying instruments such as total stations, GPS receivers, terrestrial laser scanners, and unmanned aerial vehicles (UAVs).

These instruments are used simultaneously to collect bathymetric and topographic data, resulting in a comprehensive and accurate 3D model of the surveyed area. Topo-hydrographical surveys have many important applications for integrating land-based and water-based data. One of the main uses is for engineering design, as the data collected can provide crucial information for the construction of coastal structures and infrastructure.

Additionally, these surveys are vital for coastal zone management, as they can help monitor changes in the shoreline and nearshore features over time. This is especially important in areas where there is high coastal erosion or sea level rise.

Furthermore, topo-hydrographical surveys are also used for environmental assessment, providing valuable information for coastal ecosystem conservation and management.

Hydrographical surveys are used to produce and update navigation charts, nautical maps, and electronic charting systems. These charts provide mariners and coastal authorities with crucial information to ensure safe and efficient maritime navigation. They include water depths, navigational aids, submerged hazards, and coastal features, all of which are necessary for the safe operation of ships and vessels. The responsibility for conducting these surveys and maintaining accurate navigation charts falls on hydrographical offices and maritime agencies.

They must ensure that the collected data is up-to-date and reliable in order to meet regulatory compliance and ensure the safety of those at sea. This necessitates constant monitoring and chart updating, as conditions in the waterways can change rapidly due to natural forces such as tides and storms.

Navigation charts are essential tools for all mariners, from recreational boaters to commercial shippers. They provide critical information that allows sailors to navigate safely and efficiently through harbours, shipping channels, and offshore areas. Without accurate and up-to-date charts, maritime navigation would be much more hazardous, and the risk of accidents and mishaps would increase significantly. Therefore, the production and maintenance of navigation charts through hydrographical surveys are critical for the maritime industry's safety and success.

Hydrographical and topo-hydrographical surveys are essential components of any port or harbour project. They provide critical information that is necessary for the planning, design, and construction of port infrastructure. These surveys are used to gather data on water depths, seabed conditions, tidal currents, wave heights, and sediment transport dynamics.

This information is then used to support engineering assessments, determine suitable sites, and assess the feasibility of a project. It is very important to accurately describe marine environments and coastal processes in order to make the best decisions about port layout, channel alignment, and the investments that need to be made in maritime infrastructure.

By providing detailed and precise data, surveys help ensure sustainable port development and operations. This is especially important in the current climate, where there is a growing awareness of the need for environmentally responsible practises.

With the information gathered from these surveys, engineers can make informed decisions that minimise the impact on the surrounding marine ecosystems. In addition to providing crucial information during the initial planning and design phases, surveys also play a significant role in the ongoing maintenance and management of ports and harbours.

We conduct regular surveys to monitor changes in water depths and seabed conditions, ensuring safe navigation for vessels and identifying potential hazards such as shoals or sediment buildup. This ongoing monitoring helps ensure the continued efficiency and safety of port operations by making hydrographic surveys an integral part of the entire port and harbour project lifecycle.

Surveys are essential in understanding the dynamics of coastlines and managing shoreline erosion, sedimentation, and sea-level rise.

These surveys provide valuable insights into coastal morphology, sediment transport patterns, and shoreline evolution. With this data, experts can conduct coastal vulnerability assessments, plan for beach nourishment projects, and develop strategies for coastal zone management.

Hydrographical surveys are vital for making informed decisions and implementing adaptive management strategies because they monitor coastal processes and seabed morphology. They provide critical information that helps sustainably manage coastal resources and protects communities from natural hazards and climate change effects.

With these surveys, researchers can accurately quantify the changes happening along coastlines, helping to develop long-term management plans to mitigate any potential risks. With the increasing risks posed by sea-level rise, coastal dynamics studies have become more critical than ever.

1. Maritime Services

Maritime services are an integral part of the global economy, providing a vital link between countries through the shipping industry. At their core, maritime services are responsible for facilitating the movement of goods and people across the world's oceans.

This involves a variety of activities, ranging from the management of ports and terminals to the coordination of international trade agreements. One of the key components of maritime services is the shipping industry, which plays a crucial role in the transportation of goods and materials. As the demand for goods continues to grow, the shipping industry has become increasingly important in keeping up with the pace of global trade.

The industry has developed advanced technologies and practices, such as containerization and automated cargo handling, to ensure efficient and timely delivery of goods.

In addition to facilitating trade, maritime services also play a vital role in the development and maintenance of ports and terminals. These facilities serve as hubs for the movement of goods and people and require constant management and upkeep to ensure smooth operations.

Furthermore, maritime services also encompass the coordination of international trade agreements, which are crucial in regulating the flow of goods and maintaining fair trade practises between countries. Overall, maritime services play a crucial role in keeping the global economy running smoothly and efficiently.

186

1. - Shipping

Shipping is an essential part of the modern world that has been in place for centuries. The transportation of goods and people by sea is a complex process that entails many different aspects. To ensure that this process runs smoothly, shipping services have been established. These services involve the operation of various types of vessels, each of which is designed to serve a specific purpose.

The container ship is one of the most important players in the shipping industry. These massive vessels are capable of carrying thousands of containers full of goods across the world's oceans. Container ships have revolutionised the shipping industry because they allow for more efficient and cost-effective transportation of goods. They have state-of-the-art technology and are designed to withstand the harsh conditions of the sea.

Vessels are specifically designed to transport large quantities of bulk cargo, such as grains, coal, and ore. They are equipped with various mechanisms to load, store, and unload bulk cargo efficiently. Tankers, on the other hand, are used to transport liquid cargo, such as oil and gas. These ships have special storage tanks and pumping systems to handle the transportation of such volatile

materials.

Passenger liners, as the name suggests, are used to transport people from one destination to another

Travelers often choose these ships because they provide all the amenities necessary for a comfortable journey.

187

1. - Port services

Ports are the lifeblood of the maritime industry, serving as crucial hubs where ships load and unload cargo, passengers embark and disembark, and various other activities take place. These bustling centres are responsible for customs clearances, vessel servicing, and logistics operations, making them a vital part of the global trade network. Without efficient ports, the movement of goods and people across the world would come to a grinding halt.

These ports are strategically located along major trade routes, allowing for the smooth flow of goods and services between countries. They are equipped with state-of-the-art technology and infrastructure to handle even the largest and most complex vessels, ensuring that trade can continue uninterrupted.

In addition to their economic importance, ports also play a significant role in national security. They serve as the first line of defense against potential threats with strict security measures in place, making them critical for maintaining a country's safety and security. During times of crisis, ports can also serve as emergency response centres, providing shelter and aid to those in need. Truly, ports are much more than just places to load and unload cargo; they are vital hubs that keep the world connected and moving forward.

188

1. - Maritime logistics

Maritime logistics are the backbone of global trade. Whether it's the delivery of everyday goods or the transportation of specialized cargo, maritime logistics ensures that products reach their destination in a timely and efficient manner.

This complex process involves the coordination of multiple parties, including suppliers, carriers, and customers. It is a vital component of international trade and contributes significantly to the global economy.

The management of the flow of goods, information, and financial transactions is no easy task. Maritime logistics professionals must navigate through a maze of regulations, tariffs, and customs procedures to ensure that goods are transported safely and legally. This requires a deep understanding of international trade laws and customs regulations, as well as strong communication and negotiation skills. Any mistakes in this process can have significant consequences, such as delays in delivery or even legal penalties.

Efficient maritime logistics is essential for meeting customer requirements. In today's fast-paced business environment, customers expect their orders to be delivered quickly and accurately. This puts pressure on logistics professionals to constantly improve their operations and find ways to streamline the supply chain. With the rise of e-commerce and online shopping, the demand for efficient maritime logistics has only increased.

189

1. - Security

Ensuring the safety and security of maritime activities is an important responsibility that requires a multifaceted approach. Maritime security aims to protect against various threats, including piracy, terrorism, smuggling, and illegal fishing. These threats not only endanger the lives of those involved in maritime activities but also have significant economic and environmental impacts.

As such, it is essential to have effective measures in place to prevent and respond to these threats. Patrolling is one of the most critical components of maritime security. This involves the deployment of naval vessels and aircraft to monitor and protect large areas of the ocean. Patrolling not only helps deter potential threats but also allows for a swift response to any incidents that may occur. In addition to patrolling, surveillance is critical. This can include the use of radar systems, cameras, and other monitoring equipment to detect and track suspicious activity.

Using a combination of patrolling and surveillance, maritime security forces can effectively cover vast areas of the ocean and ensure the safety of all maritime activities. International cooperation is another crucial aspect of maritime security.

As maritime activities often cross international borders, it is essential to have cooperation and coordination between different countries' security forces. This can involve intelligence sharing, joint training exercises, and coordinated response efforts. By working together, countries can strengthen their maritime security capabilities and better protect their shared waters. Overall, maritime security is a complex and ever-evolving field that requires constant vigilance and collaboration to ensure the safety and security of all maritime activities.

Law and Regulations

190

191

Maritime Law and Regulations is a comprehensive set of rules that govern the various services provided by the maritime industry. These laws and regulations are established on both international and national levels and cover a wide range of important areas. The main focus of these laws and regulations is to ensure maritime operations' safety, protect the environment, uphold labor standards, and facilitate trade activities.

On an international level, there are a number of organisations that play a crucial role in developing and implementing maritime laws and regulations. The International Maritime Organisation (IMO) is the main body responsible for promoting safety and security in international shipping and preventing marine pollution.

Additionally, the United Nations Convention on the Law of the Sea (UNCLOS) provides a framework for the settlement of disputes related to maritime boundaries and jurisdiction.

On a national level, each country has its own set of laws and regulations that apply to maritime activities within its jurisdiction. These laws can vary greatly, but they are all aimed at ensuring the smooth and safe operation of the maritime industry. They also play a key role in promoting fair trade practises and protecting the rights of maritime workers.

Compliance with these laws and regulations is essential for maintaining order and sustainability in the maritime sector.

1. - Technology

Maritime technology has moved the maritime industry into the modern era. With advancements in vessel design, navigation systems, communication tools, and automation, the industry has experienced significant improvements in efficiency, safety, and sustainability. These advancements have transformed the conduct of maritime services, enhancing their reliability and efficiency beyond measure.

The use of advanced technology has resulted in improved vessel designs that are more streamlined and fuel-efficient, reducing both operating costs and carbon emissions. In addition, the integration of navigation systems and communication tools has enhanced the safety of maritime operations.

Ships can now communicate in real-time with each other and with land-based stations, allowing for better vessel coordination and navigation. This has greatly reduced the risk of accidents at sea, making maritime services safer for both crew members and cargo.

Moreover, automation has played a significant role in improving efficiency and reducing the workload of crew members. Automation has allowed for tasks such as cargo loading and unloading, as well as maintenance and repairs, to be performed with greater precision and speed.

This has not only improved the overall efficiency of maritime services, but it has also reduced the physical strain on crew members, making their jobs safer and more manageable.

192

1. - Insurance and Finance

There are a multitude of risks associated with maritime activities, such as severe weather conditions, mechanical failures, and human error. As a result, maritime businesses have turned to financial instruments and insurance products to mitigate these risks.

These instruments include hull insurance, which protects the vessel itself; cargo insurance, which covers the goods being transported; and freight derivatives, which act as a form of insurance for the freight rates.

Hull insurance is a crucial financial instrument for maritime businesses, as it provides coverage for damage or loss to the vessel itself. This coverage extends to not only the hull of the ship but also its machinery and equipment, making it a comprehensive form of insurance.

Cargo insurance, on the other hand, protects the goods being transported by the vessel. This can include a wide range of goods, from consumer products to industrial materials. In the event of damage or loss to the cargo, the insurance company will cover the cost of the goods, providing peace of mind for both the shipper and the recipient.

In addition to these insurance products, maritime businesses also utilise freight derivatives to manage risks. These derivatives act as a form of insurance for freight rates, which can fluctuate greatly in the volatile shipping industry.

By using freight derivatives, businesses can lock in a set rate for the transportation of their goods, providing stability and predictability in their financial planning.

193

1. - Education and training

Maritime education and training are crucial in preparing individuals for their careers in the maritime industry. Training programmes and educational institutions provide the necessary skills and knowledge for a wide range of professionals, including seafarers, port workers, maritime lawyers, and logistics professionals.

These programmes offer practical and theoretical learning opportunities, enabling students to develop a deep understanding of the industry and its complexities. As they gain knowledge and experience, students are equipped to handle the challenges of the maritime world.

Maritime education and training also play a vital role in ensuring the safety and security of maritime operations. These programs train individuals on handling various vessel types, navigating through challenging weather conditions, and adhering to safety protocols.

This training is essential in preventing accidents and emergencies, which can be catastrophic in the maritime industry. Additionally, these programs also educate students on environmental sustainability, teaching them how to reduce the industry's impact on the environment. In conclusion, maritime education and training programmes are essential for individuals pursuing careers in the maritime industry.

They provide students with the necessary skills and knowledge to excel in their chosen fields and contribute to the industry's growth and development. These programmes also prioritise safety and environmental sustainability, making the maritime industry a safer and more responsible sector.

After completing their pre-sea training, aspiring seafarers are ready to take on the challenges of working onboard a ship. They have a solid foundation of knowledge and skills to build upon, enabling them to handle all kinds of situations that may arise while at sea. This

194

includes mastering navigation techniques, learning how to operate and maintain safety equipment, and understanding how to handle emergencies such as fires and medical ones.

Pre-sea training is a rigorous process that prepares individuals for a life at sea. It is not an easy feat, as it requires physical and mental strength to endure the demands of working on a ship for long periods of time. Those who successfully complete this training often have a strong sense of determination and a passion for the sea. They are dedicated to their work and are always willing to learn and improve their skills, making them valuable assets to any ship's crew.

In addition to technical skills, pre-sea training also instils important values such as teamwork, discipline, and responsibility. These are essential qualities for working on a ship, where crew members must rely on each other and work together to ensure the safety and success of the voyage.

By undergoing this training, individuals not only gain the necessary qualifications to work in the navy but also develop a strong work ethic and a sense of camaraderie that will serve them well throughout their career at sea.

Maritime Academies and Training Institutes play an important role in preparing students for a career at sea. These institutions offer comprehensive education and training programmes that are essential for aspiring seafarers.

Students at maritime academies receive training in areas such as navigation, ship handling, seamanship, and marine engineering. They also gain practical experience on board training ships, which are equipped with advanced simulation technology that simulates real-life situations at sea.

The curriculum at these academies is designed to meet the requirements of the international maritime industry. The courses are divided into different levels, with basic training being mandatory for all seafarers. This includes safety and survival training, as well as first aid and firefighting skills.

As students progress, they can choose to specialise in areas such as navigation, engineering, or cargo operations. The training is rigorous and prepares students to handle the challenges of working in the maritime industry. In addition to classroom instruction, students also participate in practical training on actual ships. This hands-on experience is crucial for developing essential skills and gaining a deeper understanding of the industry.

Many maritime academies also have partnerships with shipping companies, which provide students with internship opportunities and job placements upon graduation. With the increasing demand for skilled seafarers, maritime academies and training institutes play a vital role in shaping the future of the maritime industry and providing a pathway for individuals to pursue a fulfilling career at sea.

Cadet programmes are a great way for students to get hands-on experience in the maritime industry while still receiving the necessary classroom instruction. Maritime academies typically offer these programs, which allow students to spend time at sea as part of their training.

This provides them with the opportunity to gain practical experience under the guidance of experienced officers and crew members. During their time at sea, cadets are able to apply the knowledge they have learnt in the classroom to real-life situations. This not only helps them to better understand the material but also gives them a taste of what it's like to work in the maritime industry. It also allows them to build valuable connections and network with professionals in the field.

Cadet programmes are an essential part of preparing students for a career in the maritime profession. They provide a well-rounded education that combines both theoretical and practical training. By the time they graduate, students will have a strong foundation of knowledge and skills that will help them succeed in their future careers.

To work in the merchant navy, individuals need to obtain relevant licences and certifications issued by maritime regulatory authorities. This helps to ensure that all workers are highly qualified and skilled for their roles on the ship.

The process for obtaining these licences and certifications can be quite rigorous, as individuals must demonstrate their knowledge of maritime laws and regulations as well as their ability to handle various tasks and responsibilities on board.

For deck officers, these certifications may include a General Operator's Certificate, which allows them to operate radio equipment on board the ship, and a Radar Observer Certificate, which demonstrates their proficiency in using radar for navigation purposes.

Engineering officers may need to obtain a Certificate of Competency, which is issued after passing an examination that tests their knowledge and skills in maintaining and repairing ship engines and other mechanical systems. Ratings, on the other hand, may need to obtain a Certificate of Proficiency, which demonstrates their basic understanding of safety procedures and their ability to carry out various tasks and duties on board the ship. In order to ensure the safety and efficiency of operations at sea, it is important for all individuals working in the merchant navy to have the necessary licences and certifications. These not only provide proof of their qualifications but also serve as a standard for maintaining high levels of expertise and professionalism in the industry.

The Navy is a highly specialised field that requires rigorous training for those who wish to join its ranks. Before embarking on this journey, individuals must first complete basic training to familiarise themselves with the ins and outs of the field. However, that is just the beginning.

Aspiring seafarers have the option to pursue specialized training based on their chosen career path or area of specialization within the Navy once they complete basic training.

This will allow them to further enhance their knowledge and skills, making them more competitive in the industry. Tanker operations is one area in which individuals can receive specialised training.

Tankers are crucial for transporting large quantities of oil and other liquids across the ocean. As a result, specialised training in this area is essential for those who wish to work on these vessels.

Container shipping is another important aspect of the Merchant Navy, and individuals can receive training specifically for this type of operation. This involves learning how to safely load and unload containers, as well as how to handle them during transit.

Another area of specialisation within the Merchant Navy is offshore oil and gas operations. This type of training is ideal for those who wish to work on offshore rigs or supply vessels. It entails learning about the specific equipment and protocols used in this type of job, as well as safety procedures for working in such a high-risk environment.

Lastly, individuals can also undergo training in dynamic positioning systems, which are essential for maintaining a vessel's position and heading in adverse weather conditions. All of these training options provide individuals with the necessary skills and knowledge to excel in their chosen career path within the merchant navy.

Maritime professionals are responsible for the safety of passengers, crew, and cargo onboard. This is no easy feat, considering the vastness of the sea and the unpredictability of the weather. Therefore, continuous education and professional development are crucial for these individuals to stay up-to-date with the latest industry regulations, technological advancements, and best practises. By doing so, they can ensure that they are equipped with the necessary skills and knowledge to handle any situation that may arise while at sea. As technology advances, so too does the maritime industry. New equipment, machinery, and systems are constantly being introduced, making it essential for professionals to undergo periodic training and revalidation. This not only ensures that they are familiar with the latest technology but also helps them stay

competent and capable in their roles.

Furthermore, as regulations and safety standards constantly evolve, continuous education ensures that professionals are aware of these changes and can implement them in their daily operations.

Furthermore, continuing education and professional development play a vital role in the career progression of maritime professionals. By staying updated on industry standards and advancements, individuals can expand their skill set and take on new roles and responsibilities. This not only benefits the individual but also the company they work for, as it leads to increased efficiency and productivity. Overall, continuous education and professional development are critical for the success and growth of maritime professionals, as well as the industry as a whole.

International standards are an essential aspect of the merchant navy's training and certification programs International organizations such as the International Maritime Organization (IMO) and the International Labour Organization (ILO) create and enforce these standards.

They are responsible for setting the bar for maritime education and training across the globe, ensuring consistent and high-quality learning experiences for aspiring seafarers. The IMO and ILO work together to create and maintain international conventions and standards that govern the merchant navy's training and certification requirements. These organizations commit to equipping seafarers with the necessary skills and knowledge to operate safely and efficiently on board ships.

The standards set by the IMO and ILO cover a wide range of topics, including navigation, safety, security, and environmental protection, among others. By adhering to these standards, the merchant navy can maintain its reputation as one of the most respected and well-regulated industries in the world. Thanks to these international standards, the merchant navy can boast of having a highly skilled and well-trained workforce. These standards ensure that seafarers receive the same level of education and training, regardless of their nationality or the country they are from.

This consistency not only benefits seafarers but also the shipping industry as a whole. It promotes safe and efficient operations at sea, reduces the risk of accidents and incidents, and helps protect the marine environment.

1. - Brokers

Maritime brokers are key players in the maritime industry, serving as intermediaries between buyers and sellers in the buying, selling, and leasing of ships. They provide valuable expertise, ensuring smooth and efficient transactions while protecting the interests of both parties.

Brokers also handle the complex negotiations involved in maritime deals, making sure that all aspects of the agreement are properly addressed. These brokers are knowledgeable about the current market trends and conditions, providing valuable insights to their clients. They help find the right buyer or seller and determine the appropriate price for a vessel.

Brokers also ensure that all necessary documentation and legalities are handled properly, avoiding any future legal issues. They are also responsible for conducting thorough inspections of the ships, ensuring that they are in good condition and meet all safety standards. In addition to their role in transactions and negotiations, maritime brokers also play a crucial role in providing information and updates regarding the industry.

They keep their clients informed about any changes or developments in the market, helping them make informed decisions. Their extensive network and industry connections also allow them to connect buyers and sellers, expanding opportunities for both parties.

In short, maritime brokers are essential to keeping the maritime industry running smoothly and efficiently.

201

1. - Ship brokers

Ship brokers are vital to the maritime industry, acting as intermediaries between ship owners and charterers, or buyers and sellers of ships. They play a crucial role in facilitating the sale, purchase, and chartering of vessels of various types, including bulk carriers, tankers, container ships, and offshore vessels.

These professionals are experts in the industry, with extensive knowledge of market trends and regulations. They provide valuable market intelligence to their clients, ensuring that they make informed decisions. In addition to providing market intelligence, ship brokers also negotiate terms on behalf of their clients. This includes negotiating the price of the vessel as well as other terms such as delivery dates, payment schedules, and contract clauses.

Ship brokers are skilled negotiators who are able to find a balance between the needs of both parties and ensure a fair and mutually beneficial agreement is reached. They also assist parties in navigating the complexities of maritime transactions, which can involve various legal and logistical considerations.

Overall, ship brokers play a vital role in the smooth operation of the maritime industry. Their expertise and services are essential in ensuring that the buying, selling, and chartering of vessels is conducted efficiently and effectively. Without their assistance, the process would be much more challenging and time-consuming for both ship owners and charterers, or buyers and sellers. Ship brokers are a key component of the industry, and their contributions are essential to its success.

202

1. - Charter brokers

Charter brokers have an immense amount of knowledge that is required to make a charter contract successful. They maintain constant communication with ship owners and charterers to ensure that both parties' requirements are met. With their expertise in the shipping industry, they are able to match the needs of charterers, such as cargo owners or traders, with available vessels that meet their specific requirements in terms of size, capacity, route, and timing.

Furthermore, charter brokers play a crucial role in negotiating charter party agreements. They carefully review and analyse the terms and conditions of the contract to ensure that they are mutually beneficial for both parties. This includes ensuring that the charterer's cargo is transported safely and efficiently, while also ensuring that the ship owner's vessel is utilized to its maximum potential. Charter brokers also act as mediators in case of any disputes or issues that may arise during the duration of the charter contract.

With their extensive knowledge and experience, charter brokers are able to provide valuable insights and recommendations to both ship owners and charterers. They are constantly updated on market trends and conditions, allowing them to make informed decisions and provide the best possible solutions for their clients. Overall, charter brokers play a vital role in the shipping industry by facilitating successful charter contracts and ensuring the smooth operation of vessels.

203

1. - Freight Brokers

Freight brokers are essential in the shipping industry as they help facilitate the transportation of goods by sea. Acting as intermediaries between shippers and carriers, they play a vital role in ensuring that goods are transported safely and efficiently.

Freight brokers are known for their expertise in matching shippers with suitable vessels or shipping services for their cargo. Their knowledge and experience in negotiating freight rates and terms make them an integral part of the shipping process. Freight brokers are highly skilled professionals who are well-versed in the intricacies of the shipping industry. They often specialise in specific types of cargo or trade routes, allowing them to provide tailored solutions to their clients. With their extensive network of contacts and in-depth knowledge of the shipping market, freight brokers are able to find the best possible options for their clients.

This not only helps shippers save time and effort but also ensures that their cargo is transported in the most cost-effective manner. In today's globalized world, the demand for freight brokers is ever-increasing.

With the rise of international trade, the need for efficient and reliable transportation of goods has become paramount. This is where freight brokers play a crucial role by providing their expertise and services to facilitate the movement of goods across the globe.

204

1. - Port agent

Port agents play a vital role in the maritime industry, serving as the link between ship owners, operators, charterers, and port authorities. They are responsible for coordinating all aspects of a ship's visit to a port, from arranging for necessary services to handling important administrative tasks.

Without port agents, the process of entering and exiting a port would be chaotic and disorganised. One of the key responsibilities of port agents is to ensure the safe and efficient handling of cargo.

This includes making arrangements for pilot age, tugboat services, and berth reservations to ensure that ships can dock and unload their cargo smoothly. They also work closely with cargo handling operators to ensure that the cargo is loaded and unloaded in a timely and secure manner.

This is crucial for keeping the supply chain running smoothly and preventing delays that could have a ripple effect on the entire industry. In addition to cargo handling, port agents also play a crucial role in handling administrative tasks.

This includes dealing with customs clearance and immigration formalities, as well as managing important documentation. This is especially important in international trade, where different countries have varying regulations and requirements.

Port agents are well-versed in these procedures, and they ensure that all necessary paperwork is completed accurately and efficiently. This allows ships to enter and exit ports without any delays or issues, saving time and money for all parties involved.

205

1. - Bunker brokers

Bunker brokers are experts in the marine fuel market. They're responsible for matching ship owners and operators with the best bunker prices, as well as negotiating supply contracts with bunker suppliers. But their role goes beyond just finding the best deal. Bunker brokers also provide valuable market insights, keeping their clients informed about trends and fluctuations that might affect their business. This information is crucial for ship owners and operators, who need to make informed decisions about their fuel supply.

In addition to their knowledge of the marine fuel market, bunker brokers also help ensure that bunkers are of the highest quality. They work closely with bunker suppliers to ensure that all industry standards are met and fuel quality compliance is maintained.

This is particularly important, as poor-quality fuel can lead to serious consequences for both the ship and the environment. Bunker brokers also act as intermediaries between ship owners and operators, making sure that the fuel is delivered on time and that all parties are satisfied with the transaction. By taking care of these details, bunker brokers allow their clients to focus on other aspects of their business. Bunker brokers play a critical role in the shipping industry, bridging the gap between ship owners and bunker suppliers.

Their expertise and attention to detail ensure that their clients get the best deals on quality fuel, while they also keep them informed about market trends. By taking care of the logistics and negotiations, bunker brokers make the fuel procurement process smoother and more efficient for everyone involved.

206

1. - Insurance brokers

Insurance brokers are experts in the marine insurance industry. They are dedicated to providing the best coverage for all parties involved in maritime activities, including ship owners, charterers, cargo owners, and other stakeholders. With their knowledge and experience, they are able to accurately assess the insurance needs of their clients and help them obtain the best quotes from insurers.

They also have the skills to negotiate insurance policies that provide comprehensive protection against various risks, such as hull damage, cargo loss, liability claims, and marine perils. One of the key roles of insurance brokers is to help their clients navigate the complex world of marine insurance.

They are well-versed in the various types of coverage available and can guide their clients toward the most suitable options for their specific needs. They also act as intermediaries between their clients and insurance companies, ensuring that their clients' best interests are always represented.

This is especially important when it comes to negotiating insurance policies; brokers are able to use their expertise to secure the most favourable terms for their clients. In addition to their knowledge and expertise, insurance brokers also play a crucial role in managing risk for their clients.

They are able to identify potential risks and make recommendations on how to mitigate them, ensuring that their clients are adequately protected. This is particularly important in the maritime industry, where there are numerous risks involved in shipping and transportation. Shipowners, charterers, cargo owners, and other stakeholders can operate with confidence knowing that their insurance needs are well taken care of with the help of insurance brokers.

207

1. - Maritime legal service

Maritime legal services encompass a broad range of legal matters related to the maritime industry, shipping, and maritime trade. These services are essential for the maritime industry's smooth operation, which is an integral part of international trade. They cover a wide variety of issues, including the draughting of contracts, vessel finance, and insurance, as well as dispute resolution. The legal services provided to the maritime industry ensure that all parties involved in maritime activities are protected and their rights are safeguarded. This is especially important in cases of accidents, collisions, or other incidents at sea.

The lawyers who specialise in this field are well-versed in international maritime law as well as national and local laws, allowing them to provide comprehensive and effective legal support to their clients. In addition to the legal aspects, maritime legal services also play a crucial role in promoting safety and security at sea.

They work closely with governmental agencies and international organisations to develop and implement regulations and policies that ensure the protection of the environment and the safety of seafarers. This includes providing legal advice on issues such as pollution prevention, maritime labour laws, and compliance with international standards.

Admiralty law, also known as maritime law, is a specialised branch of law that governs all legal matters pertaining to navigation and commerce on navigable waters. This area of law is primarily concerned with regulating the conduct of individuals and businesses involved in maritime activities, including the operation of vessels, shipping contracts, and the transportation of goods and passengers.

One of the key functions of admiralty law is to provide legal protection and recourse for individuals and companies operating in

208

the maritime industry. This includes advising clients on various legal issues such as maritime contracts, vessel registration, maritime liens, salvage, towage, collision, and personal injury claims.

With the constantly evolving nature of the maritime industry, it is crucial to have a deep understanding of admiralty law in order to effectively navigate and protect the rights and interests of clients. Maritime legal services play a crucial role in ensuring the smooth and efficient operation of the global shipping industry.

1. - Maritime lawyers

Maritime lawyers specialise in vessel transactions and financing. These skilled attorneys are well-versed in the intricacies of the maritime industry and are able to assist clients in all aspects of buying, selling, and financing vessels.

They are experienced negotiators and are able to draft and review contracts and agreements that are tailored to their clients' specific needs. One of the key areas in which maritime lawyers provide their expertise is in the negotiation and draughting of contracts related to vessel sales and purchases. These agreements can be complex, and it is essential that they are carefully crafted to protect the interests of all parties involved.

Maritime lawyers are also well-versed in shipbuilding contracts, which are essential for the construction of new vessels. These contracts must be carefully reviewed to ensure that all specifications and timelines are met and that all parties are protected. In addition to vessel sales and purchases, maritime lawyers also play a vital role in ship financing. They are able to assist clients in securing mortgage agreements and ship finance arrangements, which are essential for a vessel's successful operation.

Maritime legal services are a fascinating field that requires in-depth knowledge of various international regulations and conventions. To ensure compliance with these laws, clients depend on the sound advice of maritime legal experts. These professionals assist their clients in dealing with safety concerns, environmental regulations, and labour standards.

They also advise on security, customs, and immigration requirements, providing an all-encompassing service to their clients. The importance of regulatory compliance in the maritime industry cannot be overstated. Failure to comply with these regulations can

210

lead to severe consequences, including hefty fines and criminal charges.

As such, maritime legal services play a crucial role in ensuring that all aspects of a client's business are in line with the law. By providing expert advice, these professionals help clients navigate complex regulatory frameworks, avoiding potential legal pitfalls and maintaining a good reputation in the industry. In addition to providing guidance on compliance issues, maritime legal services also assist clients in understanding the implications of these regulations on their businesses. This includes identifying potential risks and developing strategies to mitigate them.

1. - Classification societies

Maritime classification societies play a crucial role in ensuring the safety, integrity, and compliance of ships and offshore structures with international standards and regulations. These societies are independent organisations that provide classification, certification, and inspection services to the maritime industry. They are responsible for ensuring that all ships and offshore structures meet the necessary safety and environmental standards before they are allowed to operate.

These societies also play a key role in the development of new technologies and innovations in the maritime industry. They work closely with shipbuilders, designers, and other stakeholders to ensure that new vessels and structures are safe and compliant with regulations. This includes conducting research and development, testing new materials and designs, and providing guidance and recommendations for improvements.

In addition to their technical responsibilities, maritime classification societies also serve as a central hub for information and communication within the industry. They facilitate cooperation and the exchange of knowledge between different stakeholders, including shipowners, operators, governments, and regulatory bodies. This helps to ensure that all parties are on the same page and working toward a common goal of promoting safety and sustainability in the maritime sector.

The most well-known is Lloyd's Register, or LR for short, which has been around for over 260 years and is still going strong. It was founded in the United Kingdom in 1760 and has since become one of the most highly regarded classification societies in the world. LR's main focus is on providing services such as certification, classification, and technical consultancy for ships, offshore installations, and marine equipment.

212

However, it is best known for its exceptional expertise in the areas of ship classification and risk management. LR's long-standing reputation as one of the top classification societies in the world is a testament to its dedication to ensuring the safety and reliability of ships and marine equipment. Its classification services involve inspecting and certifying vessels to ensure they meet certain standards of safety and quality. This is crucial in the maritime industry, where even the smallest error or oversight can have disastrous consequences.

LR's thorough and rigorous approach to classification has earned it the trust and respect of ship owners, operators, and regulators worldwide. But LR's expertise goes beyond just ship classification.

The organisation also provides technical consulting services, offering advice and guidance on complex technical issues related to ships and offshore installations. This is a valuable resource for companies and individuals in the maritime industry who are looking to ensure the efficiency and safety of their operations. With its long history and wealth of knowledge, it's no wonder that LR continues to be a leading authority in the world of maritime classification and consultancy.

1. - Organisations and associations

Maritime organisations and associations play vital roles in advocating for the interests of various stakeholders within the maritime industry, promoting safety, sustainability, and cooperation, and facilitating information exchange and networking. These organisations act as a bridge between the various players in the industry, bringing them together to work toward common goals and address challenges that affect the industry as a whole.

By providing a platform for communication and collaboration, they ensure that the maritime industry operates as a cohesive unit with a shared vision for a safer and more sustainable future. The International Maritime Organisation (IMO) is one such organisation that has been instrumental in shaping the global maritime industry. As a specialised agency of the United Nations, it is responsible for regulating international shipping, developing and maintaining international conventions and regulations on safety, security, environmental protection, and maritime transportation.

The IMO's efforts have led to significant improvements in the safety and sustainability of the industry, making it a crucial player in the global maritime landscape. Apart from the IMO, there are several other influential organisations that contribute to the development and growth of the maritime industry.

The International Chamber of Shipping (ICS) is the principal international trade association for shipowners and operators, representing national shipowners' associations from around the world. It works toward promoting the interests of the global shipping industry by advocating for issues such as regulation, trade facilitation, safety, and environmental protection.

214

These organizations, along with others like the International Association of Classification Societies (IACS) and the International Transport Workers' Federation (ITF), have a significant impact on the maritime industry, shaping its future and ensuring its sustainability.

1. - Consultants and inspectors

Maritime consultants and inspectors are vital players in the maritime industry, providing specialised services to stakeholders. They offer extensive support in terms of compliance and risk management, ensuring that any possible issues are mitigated and handled promptly.

These professionals are equipped with specific knowledge and expertise, which allows them to provide tailored solutions to the complex challenges faced by the maritime industry. Their services are in high demand due to their ability to assist stakeholders in improving their operations.

Maritime consultants and inspectors work closely with their clients to identify inefficiencies and provide recommendations to ensure smooth operations. Their extensive knowledge of industry standards, regulations, and best practises allows them to assess the existing processes and suggest improvements, which ultimately leads to increased efficiency.

With their support, stakeholders can focus on their core business activities while ensuring that their operations meet industry standards. Their presence helps in maintaining a healthy and competitive environment in the maritime industry. Working with various stakeholders, maritime consultants and inspectors play a crucial role in ensuring that all players abide by industry regulations and standards.

216

1. - Marine surveyors

Marine surveyors are a vital part of the maritime industry, carrying out inspections, surveys, and assessments of ships, vessels, and marine equipment. Their role is to ensure that these are in compliance with the required regulatory standards, safety requirements, and industry best practises.

The job of a marine surveyor is to carefully examine the different components of a ship, such as the hulls, machinery, navigation equipment, and cargo handling systems, to ensure that they meet the necessary standards for seaworthiness and operational integrity. These professionals are well-trained and highly skilled in their field. They must have a strong understanding of maritime laws and regulations, as well as knowledge of ship construction, repair, and maintenance. They also use specialised equipment and techniques to conduct their inspections, such as ultrasonic testing and thermal imaging.

Marine surveyors play a crucial role in ensuring the safety of ships, the protection of the marine environment, and the smooth operation of the maritime industry. With their expertise and attention to detail, marine surveyors help to prevent accidents, ensure compliance with regulations, and identify potential issues before they become major problems. They are also responsible for providing detailed reports and recommendations to ship owners and operators, which can help them make necessary repairs and improvements to their vessels.

217

1. - Maritime Engineering Consultants

Maritime Engineering Consultants offer a wide range of services to their clients, including naval architecture, marine engineering, structural analysis, and offshore engineering.

These experts are highly skilled in their fields and work closely with their clients to provide the best possible solutions to their problems. They are able to offer design, analysis, and optimisation services for various types of marine structures, including ships, offshore structures, port facilities, and marine infrastructure projects. With their extensive knowledge and experience, they are able to provide innovative and efficient solutions to complex engineering problems.

One of the main areas of expertise for maritime engineering consultants is naval architecture. This involves the design and construction of ships and other marine vessels. These consultants work closely with shipbuilders to ensure that the vessels are designed to be safe, efficient, and cost-effective. They use advanced software and simulation tools to analyze the ships' performance and make necessary changes to improve their design. In addition to naval architecture, maritime engineering consultants also specialise in marine engineering. This involves the design, construction, and maintenance of various marine structures, such as offshore platforms, pipelines, and ports.

These consultants work closely with their clients to ensure that these structures are designed to withstand harsh marine environments and meet strict safety standards. They also provide valuable expertise in areas such as structural analysis and offshore engineering, ensuring that their clients' projects are completed successfully and efficiently.

218

1. - Safety Consultants

Maritime safety consultants are professionals who have expertise in evaluating and enhancing the safety management of various maritime organisations. They are especially skilled in assessing and improving safety management systems, emergency preparedness, and risk mitigation measures. In order to achieve this, they conduct a range of activities, such as safety audits, risk assessments, and training programmes.

These initiatives aim to promote a strong safety culture and ensure compliance with all relevant safety regulations. One of the key responsibilities of maritime safety consultants is to conduct regular safety audits. These audits involve thorough inspections of the organization's safety systems and procedures, identifying any gaps or weaknesses that may exist. Based on their findings, consultants provide recommendations and solutions to improve the safety management system. In addition to audits, maritime safety consultants also conduct risk assessments to identify potential hazards and develop strategies to mitigate them.

This helps organisations be better prepared for any potential emergencies. Training is another important aspect of maritime safety consultants' work. They design and deliver comprehensive training programmes to educate employees on safety procedures, emergency protocols, and risk management strategies.

These training programs not only strengthen the organization's safety culture but also guarantee that all employees possess the essential knowledge and skills to handle emergency situations.

219

1. - Environmental Consultants

Environmental consultants play a crucial role in the maritime industry. They are highly trained professionals who help ensure that the industry operates in an environmentally responsible manner. Their expertise lies in environmental management, pollution prevention, and sustainability initiatives.

By working closely with shipowners, operators, and port authorities, they assist in the implementation of environmental management systems, the development of pollution response plans, and compliance with environmental regulations.

One of the key responsibilities of environmental consultants is to help clients navigate the complex web of environmental regulations. This involves staying up-to-date on the latest laws and regulations pertaining to the maritime industry and advising clients on how to comply with them.

They also work closely with clients to develop and implement strategies for reducing their environmental impact. This may include identifying and addressing potential sources of pollution, implementing sustainable practises, and finding ways to reduce energy consumption. In addition to their technical expertise, environmental consultants also play an important role in raising awareness about environmental issues within the maritime industry. They work to educate clients on the importance of environmental stewardship and help them understand the potential consequences of not being environmentally responsible.

220

1. - Security consultants

Maritime security consultants offer a wide range of services to combat security threats in the maritime domain. With recent piracy and terrorist incidents, their services have become increasingly important.

These experts conduct thorough security assessments, identifying any potential risks and vulnerabilities. This allows them to develop customised security plans for ships and ports to ensure the safety of personnel and cargo. In addition to risk mitigation, maritime security consultants also provide training to ship crews and port personnel. This training focuses on enhancing security awareness and preparedness, empowering individuals to handle potential threats with confidence.

Through this training, individuals are equipped with the necessary skills and knowledge to react quickly and effectively in the event of an emergency. The services offered by maritime security consultants are crucial to maintaining safe and secure operations in the maritime industry.

With their expertise and experience, they play a vital role in protecting ships and ports from security threats.

221

1. - Logistics and Supply Chain Consultants

Consultants in logistics and supply chain management are highly skilled professionals committed to optimizing maritime logistics and supply chain operations. They are able to significantly increase the efficiency of these processes while reducing costs for their clients.

This is essential for businesses that are looking to remain competitive in today's fast-paced market. These consultants have a deep understanding of various areas, including cargo handling, freight forwarding, warehousing, inventory management, and transport planning.

With their expertise, they are able to provide valuable insights and recommendations that can greatly benefit their clients. One of the key roles of a logistics and supply chain consultant is to identify areas within a company's logistics and supply chain processes that can be improved. By analysing these processes, they are able to identify inefficiencies and suggest solutions that can streamline operations and reduce costs. This often involves the implementation of new technologies or processes that are designed to improve efficiency and accuracy.

By doing so, businesses are able to save time and money while becoming more competitive in their respective industries. Furthermore, logistics and supply chain consultants play a crucial role in helping businesses adapt to ever-changing market conditions. With their expertise, they are able to analyse market trends and provide recommendations that can help businesses prepare for potential challenges or opportunities. This allows companies to stay ahead of the curve and maintain a competitive edge in their industry. Ultimately, the services provided by logistics and supply chain

222

223

consultants are essential for businesses looking to optimize their operations and remain competitive in today's global market.

1. - Legal and regulatory consultants

Legal and regulatory consultants play a vital role in the maritime industry. Their expertise in maritime law, regulations, and compliance standards guarantees strict adherence to all laws and regulations.

Their job is to provide legal advice and regulatory analysis to businesses operating in the maritime sector. These consultants have extensive knowledge of the legal framework that governs the maritime industry and help companies comply with the rules and regulations.

They also offer assistance with contractual disputes and can even represent their clients in maritime litigation. The maritime industry is highly regulated due to the risks involved in the transportation of goods and people across the seas. Therefore, it is crucial for businesses to have legal and regulatory consultants on their side to guide them through the complex legal landscape. These consultants understand the intricacies of maritime laws and regulations and can help companies navigate through them smoothly.

They also keep businesses updated on any changes or updates in the laws, ensuring that their clients are always in compliance.

224

1. - Port and Terminal Consultants

Port and terminal consultants are professionals who assist with the optimisation of port operations. They provide services such as infrastructure development, management strategies, and port planning.

Their expertise also extends to areas such as berth optimisation, terminal design, cargo handling efficiency, and port performance improvement. With their assistance, ports can operate more smoothly and efficiently, ensuring that cargo and other goods are handled in a timely and cost-effective manner.

These consultants are highly experienced in the field of ports and terminals, and they possess the knowledge and skills to identify areas for improvement. They work closely with port authorities and stakeholders to provide solutions that are tailored to each port's specific needs. With their expertise, they can analyse and assess the current state of port operations and develop strategies to enhance their performance.

This includes optimising the layout of berths, improving cargo handling processes, and implementing management strategies that can increase efficiency and reduce costs. In addition to their technical knowledge, port and terminal consultants also have a deep understanding of the industry and its regulations.

They are familiar with the latest trends and technologies in port operations and can provide guidance on how to implement them effectively. By working with these consultants, port authorities can stay ahead of the competition and maintain a competitive edge.

225

14 - Maritime Information and Communications Technology

Maritime information and communications technology (ICT) services are essential for the success of the global maritime industry. These services include a variety of technologies and solutions that are designed to improve the efficiency, safety, and connectivity of maritime operations.

As technology advances rapidly, maritime companies are increasingly relying on ICT services to improve their operations and stay competitive in the industry. One of the key benefits of maritime ICT services is the improvement in efficiency.

With advanced technologies such as automation and data analytics, maritime companies can streamline their processes and eliminate time-consuming tasks. This ultimately leads to increased productivity and cost savings. For instance, the use of ICT services has enabled shipping companies to optimise their routes, reduce fuel consumption, and improve vessel maintenance, resulting in significant cost savings.

Moreover, maritime ICT services also play a crucial role in improving safety within the industry. With the help of technologies such as satellite tracking and real-time monitoring, maritime companies can ensure the safety of their vessels and crew members. These services also provide early warning systems for potential hazards, allowing companies to take the necessary precautions and prevent accidents. Furthermore, the use of ICT services has improved the connectivity between vessels and onshore operations, enabling better communication and coordination for effective decision-making.

There has been a growing need for the shipping industry to stay connected while out at sea. Maritime ICT services have stepped

**226

up to meet that need by providing connectivity solutions that cater to the maritime environment's unique requirements. These services offer a wide range of options, such as satellite communication systems, VSAT (Very Small Aperture Terminal) services, maritime broadband, and mobile networks specifically optimised for maritime use. This ensures that ships and vessels are able to stay connected, no matter how remote their location.

Satellite communication systems are one of the most important aspects of maritime connectivity solutions. These systems utilize satellites to establish communication links between ships and shore stations, ensuring that distance or geographical barriers do not hinder communication. In addition, VSAT services provide high-speed and reliable internet connections, allowing for seamless communication and data transfer.

Maritime broadband further enhances these capabilities by providing a larger bandwidth and faster speeds, enabling ships to access a wide range of applications and services. Another important feature of maritime ICT services is the optimisation of mobile networks for maritime use. These networks are specifically designed to withstand the challenges of the open sea, such as strong winds, rough weather, and saltwater exposure.

14.1 - Shipboard Communication Systems

Seamless communication is critical to the success of any sea voyage. Shipboard communication systems provide a convenient and efficient way for crew members to communicate with each other, as well as with vessels and shore-based operations. With these systems, crew members can easily relay important information to one another, such as safety protocols or course adjustments.

Voice communication is one of the most important components of shipboard communication systems. This allows crew members to speak to each other directly, whether they are on the same vessel or different ones. In addition, data transmission allows for the transfer of important data, such as weather reports or navigational information, between vessels and shore-based operations.

This ensures that everyone involved in the voyage is on the same page and can make informed decisions. Shipboard communication systems also provide access to email services and the internet.

It allows crew members to stay connected with loved ones back home and access important resources and information while at sea. This can be especially crucial in emergency situations, where quick and reliable communication can be the difference between life and death.

228

14.2 - Vessel Tracking and Monitoring

Maritime ICT services are used for vessel tracking and monitoring, enabling operators to obtain real-time information on vessel locations, routes, and speed. They guarantee the capture and analysis of all crucial data, facilitating improved decision-making and heightened safety.

These services use a variety of technologies, including the Automatic Identification System (AIS) and satellite tracking, to provide a comprehensive view of vessel operations. One of the key benefits of vessel tracking and monitoring is enhanced situational awareness.

With real-time data, operators have a complete picture of vessel movements, enabling them to make informed decisions in a timely manner. This is particularly crucial in emergency situations, where quick and accurate information is vital for ensuring the safety of both the vessel and its crew.

These services also provide valuable insights into the efficiency of vessel operations, allowing for optimisations that can help reduce costs and improve overall performance. In addition to vessel tracking and monitoring, maritime ICT services also offer solutions for remote monitoring and control of vessel systems. This includes monitoring engine performance, fuel consumption, and other critical parameters. With this information, operators can proactively identify and address potential issues, minimising downtime and maintenance costs.

229

1. - The maritime Internet of Things

The maritime industry is rapidly adopting IoT technologies. These technologies collect data from onboard sensors, equipment, and systems to monitor and optimize various aspects of vessel operations. By integrating IoT devices with data analytics platforms, maritime ICT services enable vessel owners and operators to improve their vessel's performance, fuel efficiency, and maintenance planning.

One of the key benefits of using IoT in the maritime industry is the ability to collect and analyse real-time data from various sources. This data can provide valuable insights into the operational performance of a vessel, allowing for more informed decision-making. For example, by monitoring fuel consumption and engine performance, ship owners can identify areas for improvement and implement strategies to increase fuel efficiency.

Additionally, by tracking equipment and machinery, maritime companies can proactively schedule maintenance and repairs, reducing downtime and costly breakdowns. In conclusion, the use of IoT technologies in the maritime industry is revolutionising vessel operations.

By integrating IoT devices and data analytics platforms, maritime ICT services provide valuable insights and opportunities for optimisation. With the ability to collect and analyse real-time data, vessel owners and operators can improve performance, reduce costs, and enhance overall efficiency.

1. - Maritime Safety and Navigation Systems

Maritime Safety and Navigation Systems are extremely important when it comes to keeping our shipping lanes safe and reliable. With the use of ICT services, we can continue to enhance these systems and help ensure that ships are able to navigate safely and efficiently. The technology used in these systems has advanced

230

231

over the years and continues to improve, thanks to the development and deployment of advanced navigation systems. As ships are required to travel through challenging conditions, these advanced systems have been designed to help mitigate risks and improve vessel navigation.

Electronic charting has greatly improved navigation accuracy. Collision avoidance systems are now able to accurately detect and avoid potential collisions with other ships and dangerous obstacles. Furthermore, the use of weather routing services has allowed ships to better navigate around storms and severe weather.

The use of ICT services to develop and deploy advanced navigation systems has created a safer environment for vessels, increasing shipping efficiency.

14.5 - Port and Terminal Management Systems

Port and Terminal Management Systems are an essential part of port operations, cargo handling, and logistics processes. The use of ICT solutions in these systems has greatly improved the port industry's efficiency and productivity. These systems are designed to streamline and automate various tasks, such as vessel scheduling, cargo tracking, and inventory management.

With the integration of port management software, terminal operating systems, container tracking systems, and electronic data interchange platforms, port and terminal management have become more efficient and seamless.

One of the major benefits of these systems is their ability to exchange information between stakeholders in real time. This has greatly improved communication and collaboration among different parties involved in port and terminal operations. Electronic data interchange allows stakeholders to easily share information such as vessel schedules, cargo manifests, and customs documents, resulting in faster clearance and turnaround times.

Additionally, these systems also provide real-time tracking and monitoring of cargo, which helps reduce the risk of theft and damage. The growth of global trade has led to an increase in demand for efficient and advanced port and terminal management systems in recent years.

Ports and terminals are under increasing pressure to handle larger cargo volumes while maintaining high safety and security standards. With the continuous development and integration of new technologies, these systems are constantly evolving to meet the changing demands of the industry.

232

14.6 - Maritime Cyber security

Maritime cyber security is a rapidly growing concern as the maritime sector continues to embrace digitisation. This has led to an increase in demand for ICT services that offer tailored cyber security solutions.

These services cater to the diverse needs of the maritime industry, providing network security, endpoint protection, and intrusion detection systems. In addition, there is now a growing demand for cyber security training for maritime personnel, which has become increasingly necessary to keep up with the evolving cyber security threats.

The maritime industry's need for cyber security has become more pressing as it relies heavily on technology for its operations. With the rise of cyber attacks and the potential consequences they can have on the safety and security of vessels, cargo, and crew, it has become essential to have robust cyber security measures in place. Maritime ICT services understand the unique challenges faced by the industry and provide specialised solutions to protect against cyber threats. In addition to providing cyber security solutions, maritime ICT services also play a critical role in raising awareness and educating maritime personnel on cyber security best practises. This includes training on how to identify and respond to potential cyber threats, as well as implementing protocols to prevent and mitigate attacks. With the continuous evolution of technology, it is important for the maritime industry to stay updated and proactive in protecting its operations from cyber attacks.

233

14.7 - Remote Monitoring and Maintenance

With the advancement of technology, ICT services have allowed for the remote monitoring and maintenance of onboard systems and equipment. These services utilize remote diagnostic tools, predictive analytics, and condition monitoring solutions to provide a comprehensive view of a vessel's performance. This not only allows for early detection of potential issues but also enables proactive maintenance planning, minimising downtime, and optimising vessel performance.

Through remote monitoring and maintenance, vessel operators can effectively manage their fleet without the need for a physical presence on board. These services also allow for real-time monitoring of critical systems, providing a more immediate response to any potential issues.

In addition, the use of remote diagnostic tools and predictive analytics also allows for more accurate and timely maintenance planning. By constantly monitoring the condition of onboard systems, potential issues can be identified and addressed before they become major problems.

Electronic Documentation and Compliance

With the introduction of electronic logbooks, electronic chart display and information systems (ECDIS), and electronic recordkeeping systems, ships can now easily comply with international regulations.

These services have digitised the entire documentation process, making it more efficient and accurate. Electronic logbooks have become an essential tool for ships, as they eliminate the need for manual data entry and reduce the risk of human error. This not only saves time but also ensures compliance with regulations. Similarly,

234

electronic chart display and information systems (ECDIS) have replaced traditional paper charts, providing ships with real-time navigational data and improving safety at sea.

These advanced systems allow ships to easily comply with international navigation and route planning regulations. Moreover, electronic recordkeeping systems have made compliance with international regulations a seamless process. These systems enable ships to store and access important documents electronically, eliminating the need for physical storage space and reducing the risk of loss or damage.

15 - Maritime protection and indemnity

Maritime Protection & Indemnity (P&I) clubs are a vital part of the maritime industry. These clubs are mutual insurance associations that provide both liability coverage and risk management services to shipowners, operators, and charterers.

They are essential for the industry's smooth operation and ensure that all parties involved are protected from potential financial risks. These clubs not only offer insurance coverage but also provide risk management services.

This includes giving advice on how to prevent losses and minimise risks, as well as providing support in the event of a maritime incident. This support can range from legal assistance to financial aid, depending on the severity of the incident.

P&I clubs play a crucial role in keeping the maritime industry safe and secure, as well as ensuring that any potential losses are covered. In addition to their insurance and risk management services, P&I clubs also contribute to the overall development and growth of the maritime industry.

P&I clubs are responsible for liability insurance coverage for all the risks that shipowners may encounter in their operations. This includes all third-party liabilities such as pollution, collisions, personal injury, cargo damage, and wreck removal costs. This insurance coverage is important because it protects from the financial burden of these potential risks and liabilities, which can be extremely costly.

One of the main benefits of P&I club liability coverage is that it offers a wide range of coverage for different types of risks. This ensures that they are protected from a variety of potential liabilities and are not just limited to one type of risk.

: Furthermore, the coverage for third-party liabilities is essential as it safeguards not only the shipowners but also any third parties

236

impacted by the ship's operations, including cargo owners or potentially injured individuals.

They operate on a mutual basis, which means that their members work together to provide insurance coverage and share risks. This mutual structure is beneficial for both the members and the club, as it helps to spread the risk and reduce the financial burden on individual members.

One of the key features of P&I clubs is that members pay annual premiums based on the size and type of their vessels. This customization of premiums to each member's individual needs and risks makes it a fair and cost-effective insurance option.

However, in the event of any major claims, members may be required to contribute additional funds to cover the costs. This ensures that the club has enough resources to handle any unforeseen events and provide timely and adequate compensation to its members.

Overall, the mutual insurance structure of P&I clubs promotes a sense of community and cooperation within the maritime industry. It enables members to share knowledge and resources and establishes a strong support system for all participants. Moreover, the constant review and assessment of risks and premiums ensure that the club remains financially stable and is able to provide reliable insurance coverage for its members.

P&I clubs, or protection and indemnity clubs, are specialised insurance groups that offer services beyond traditional insurance policies. In addition to financial protection, these clubs provide risk management services to their members.

By seeking to mitigate and prevent liabilities, P&I clubs keep their members informed of safety management systems and best practises for handling cargo. This includes ensuring compliance with international regulations and assisting with emergency response planning. Through their risk management services, P&I clubs aim to provide their members with a more complete picture of potential risks and how to prevent them.

By offering guidance and resources, they help members stay on track and transition smoothly in the event of an emergency. With the support and expertise provided by P&I clubs, members can effectively manage risks and minimise potential liabilities. This ultimately benefits both the members and the overall industry by promoting safer practises and reducing the potential for accidents and incidents.

They have dedicated claims teams, each with their own expertise in handling claims on behalf of their members. These teams are made up of experienced maritime lawyers, marine surveyors, and claims handlers who are well-versed in the complex world of maritime law.

They work tirelessly to ensure that their members' interests are protected and that claims are resolved as quickly and smoothly as possible. When a claim is filed against a member, the club's claims team immediately begins their investigation. This involves gathering evidence, interviewing witnesses, and consulting with experts to determine the cause and extent of the incident. Once liability has been established, the team then works to negotiate a fair settlement with the claimant. This process can be lengthy and complicated, but the goal is always to protect the member's financial interests and minimise any potential losses.

The P&I clubs provide emergency response services to members who are dealing with a crisis. These services include pollution incidents, salvage operations, and casualty response. Their main goal is to offer expertise, resources, and coordination support to help members minimise the impact of emergencies and protect their interests. This can be especially helpful in situations where quick and decisive action is necessary. In the event of a pollution incident, P&I clubs can provide members with the necessary resources and knowledge to properly contain and clean up the pollution.

They can also assist with legal issues and ensure that members are in compliance with any applicable environmental regulations. For salvage operations, P&I clubs can provide expert advice and coordinate with salvage companies to ensure that it is carried out safely and efficiently. In cases of casualty response, they can help with the management and coordination of resources, such as tugboats and divers, to minimise the risks and damages.

P&I clubs play a crucial role in the maritime industry because they provide financial stability and security to their members P&I clubs achieve this through various means, including reinsurance arrangements, reserve funds, and prudent financial management practices. These measures are in place to ensure that the clubs can fulfil their obligations in the event of large or catastrophic claims.

This not only benefits the members but also the entire industry by promoting a sense of security and trust. Implementing effective risk management practises is one of the key aspects of financial stability and security. : Effective risk management practices must be implemented P&I clubs constantly monitor and assess potential risks in order to mitigate them and maintain their financial stability. This includes identifying potential hazards, evaluating their impact, and taking the necessary measures to minimise any potential losses.

They are able to safeguard their financial position and ensure that they meet their obligations to their members. Moreover, the financial stability and security provided by P&I clubs are not only limited to their members but also extend to other stakeholders such as shipping companies, ports, and cargo owners.

To protect their own interests, these parties rely on the clubs' financial stability. This highlights the importance of P&I clubs in the maritime industry and reinforces the need for them to maintain strong financial stability and security.

16 - Strategic Thinking

The importance of strategic thinking cannot be overstated when it comes to the maritime industry. In today's ever-evolving world, economies heavily reliant on maritime trade and commerce must constantly adapt and innovate to stay competitive. This requires a forward-thinking mindset that is always looking for ways to improve and grow.

By embracing innovation and proactively addressing challenges and opportunities, economies can position themselves for long-term success in the maritime industry. One of the key benefits of strategic thinking in the maritime industry is the ability to stay ahead of the curve.

By anticipating potential challenges and opportunities, economies can better prepare for and respond to changes in the industry. This not only allows for smoother operations but also creates a more stable and resilient economy. In addition, strategic thinking encourages a culture of continuous improvement and growth, which can lead to increased efficiency and profitability.

However, strategic thinking is not just about reacting to external factors. It also involves actively seeking out new opportunities and ways to improve. By constantly evaluating and adjusting strategies, economies can stay competitive and thrive in the dynamic and interconnected world of maritime trade and commerce. This proactive approach can also lead to the development of new technologies and practises that can benefit the entire industry. In short, strategic thinking is crucial for the survival and development of economies reliant on the maritime industry.

240

16.1 - Navigating Uncertainty

The maritime industry is not only an exciting career but also a challenging one. With ever-changing global trade, political tensions, and natural disasters, shipping companies are constantly under threat. Even with the best-laid plans, the industry is subject to numerous uncertainties, and therefore, stakeholders must be strategic in their approach. This is where strategic thinking comes in, empowering those in the industry to anticipate and prepare for the unknown.

Stakeholders who have a strategic mindset can develop contingency plans to mitigate the risks posed by uncertainties. With a clear understanding of the potential challenges, companies can adapt to changing circumstances effectively, ensuring the smooth operation of their business.

Strategic thinking also helps in identifying new opportunities that may arise from these uncertainties, providing a competitive advantage for companies in the maritime industry. In order to navigate through the sea of uncertainties, stakeholders must be constantly evolving and adapting to new situations.

This requires not only strategic thinking but also the ability to envision different scenarios and make informed decisions based on them. It is a crucial aspect of the maritime industry, and without it, companies may struggle to survive in such a volatile and unpredictable environment. By staying on track and transitioning smoothly, stakeholders can successfully navigate through these uncertainties and come out stronger on the other side.

241

16.2 - Capitalizing on Opportunities

With strategic thinking, businesses can identify and seize emerging opportunities in the maritime industry to foster economic growth. This can involve venturing into untapped markets, broadening their range of services, adopting cutting-edge technologies, or forming strategic alliances to gain an edge over competitors.

By capitalising on these opportunities, companies can expand their reach and increase their market share, leading to improved financial performance. Moreover, strategic thinking equips organisations with the foresight to anticipate future trends and potential challenges.

By staying ahead of the curve, they can proactively plan and adapt their strategies to capitalise on upcoming opportunities and mitigate risks. This not only helps them stay competitive but also positions them as industry leaders, paving the way for long-term success. In conclusion, strategic thinking is crucial for businesses operating in the maritime sector.

By leveraging this mindset, companies can uncover and capitalise on emerging opportunities, drive economic growth, and position themselves for sustained success in the dynamic maritime industry. With the right approach, organisations can stay ahead of the competition and thrive in an ever-changing business landscape.

242

16.3 - Managing Risks

Strategic thinking involves assessing and managing the risks inherent in maritime operations.

Operational risks are those that arise from daily operations and have the potential to impact a maritime venture's success. Financial risks include the potential for financial losses due to market fluctuations, currency exchange rates, and economic instability.

Legal risks refer to the potential for legal action or penalties resulting from non-compliance with laws and regulations.

Reputational risks are caused by negative publicity or actions that damage a country's image or reputation. Proactively identifying and mitigating risks is crucial for safeguarding maritime interests and minimising potential losses. This can involve conducting thorough risk assessments, implementing risk management strategies, and continuously monitoring for potential risks.

By doing so, economies can better protect their maritime operations and investments and also strengthen their position in the global market. Failure to manage risks effectively can result in significant financial losses, legal consequences, and reputational damage, which can have far-reaching consequences for a country's economy. Strategic thinking and risk management go hand in hand when it comes to the maritime industry.

It is essential for economies to have a comprehensive understanding of the various risks associated with maritime operations and take proactive measures to address them.

490

243

16.4 - Enhancing Competitiveness

Strategic thinking is an important aspect of the success of any organisation, especially those involved in the maritime industry. It plays a crucial role in shaping the future of these industries as they strive to remain competitive in the global market.

This involves creating a long-term vision and setting strategic goals to achieve it. With the ever-changing market landscape, it is essential for maritime industries to adopt a strategic mindset in order to stay ahead of the curve. In addition to setting goals and establishing a vision, strategic thinking also involves the implementation of various initiatives.

These initiatives aim to improve efficiency, productivity, and innovation within maritime operations, infrastructure, and services. By constantly seeking ways to enhance these aspects, maritime industries are able to keep up with the market's demands and stay relevant.

This also allows them to adapt to any changes in the industry, ensuring their continued success. In conclusion, strategic thinking is crucial for the growth and competitiveness of maritime industries on a global scale. It enables these industries to stay ahead of the competition by developing a long-term vision, setting strategic goals, and implementing initiatives to improve efficiency, productivity, and innovation. Strategic thinking is essential in order to stay relevant and successful in the maritime industry, given the constantly evolving market.

244

16.5 - Promoting Sustainability

In order to promote sustainability in the maritime sector, strategic thinking is of utmost importance. This type of thinking is critical for addressing the industry's various environmental challenges and mitigating its impact on ecosystems and communities.

One of the key ways to achieve this is through the adoption of green technologies. By implementing these technologies, the industry can significantly reduce its carbon footprint and help preserve the environment.

Furthermore, implementing sustainable practises, such as waste reduction and resource conservation, is essential to promoting sustainability in the maritime sector.

Another crucial aspect of promoting sustainability in the maritime sector is the support of regulatory frameworks that prioritise environmental stewardship. These frameworks can facilitate the industry's accountability for its environmental impact and its efforts to mitigate any adverse effects.

By working together with governments and regulatory bodies, the maritime industry can play a significant role in protecting the environment and promoting sustainable development. This requires a collaborative effort and a commitment to prioritise environmental concerns. In conclusion, promoting sustainability in the maritime sector is a complex and multifaceted task that requires strategic thinking, the adoption of green technologies, and the support of regulatory frameworks. By implementing sustainable practises, the industry can mitigate its impact on the environment and contribute to the development of a more sustainable future.

245

16.6 - Fostering Collaboration

Strategic thinking is an essential element in fostering collaboration among stakeholders in the maritime ecosystem. It encourages cooperation and helps to build partnerships between governments, industry players, academia, and civil society.

By bringing together a diverse set of perspectives and expertise, strategic thinking enables economies to leverage collective resources and networks to address common challenges and achieve shared objectives. This collaborative approach not only enhances efficiency and effectiveness but also fosters innovation and creativity.

Moreover, strategic thinking promotes a culture of inclusivity and open communication, which are crucial for successful collaboration. By encouraging stakeholders to share their unique insights and perspectives, it helps to create a more comprehensive understanding of complex issues and enables the development of more robust solutions.

This collaborative approach also facilitates the identification and mitigation of potential conflicts of interest, ensuring that all stakeholders' interests are considered and addressed.

246

16.7 - Adapting to technological innovation

The maritime industry is an essential part of the global economy. It handles the majority of the world's trade, making it vital for its continued growth and development.

However, for this industry to remain relevant, it must adapt to the ever-changing technological landscape. That's why several companies and organizations are investing in new technologies to transform the maritime sector.

These innovations, such as autonomous vessels, digitalisation, block chain, and artificial intelligence, are driving the industry towards a more efficient and sustainable future. One of the primary reasons for this technological transformation is the need to increase efficiency and reduce costs.

Traditional shipping methods have been plagued by delays, errors, and inefficiencies, resulting in significant financial losses. These issues can be significantly reduced, if not eliminated, with the introduction of new technologies. Autonomous vessels can improve navigation and reduce human error, while block chain technology can streamline supply chain processes and reduce paperwork. These advancements will not only save time and money but also improve the overall safety and reliability of the maritime industry.

Adapting to technological innovation is not without its challenges. Companies must be strategic in their approach and carefully consider the potential risks and benefits. For instance, while automation and digitalisation can improve efficiency, they can also lead to job displacement. Therefore, it is crucial to prepare for the future of maritime transportation and logistics by upskilling and rescaling the workforce.

247

1. - Maritime Environment and Nature Conservation

The maritime environment and nature conservation are essential aspects of sustainable maritime development. This ensures the preservation of marine ecosystems, biodiversity, and natural resources for present and future generations.

The maritime environment plays an important role in ecosystem restoration. It also supports the livelihoods of coastal communities and provides food security for millions of people around the world. As a result, it is critical to protect and conserve the maritime environment for both humans and the planet's well-being.

Nature conservation is another important aspect of sustainable maritime development. It involves protecting and managing natural resources such as forests, oceans, and wildlife. This is necessary to maintain ecosystem balance and prevent activities that may harm the environment. It also ensures that natural resources are used in a sustainable manner without depleting them for future generations.

By preserving nature, we can also mitigate the effects of climate change and maintain a healthy environment. Maritime environment and nature conservation work hand in hand to achieve sustainable maritime development.

Protecting marine ecosystems and conserving natural resources can ensure the long-term health of the oceans and the communities that depend on them. This not only benefits our present generation but also secures a better future.

Marine Protected Areas (MPAs)

Marine Protected Areas (MPAs): MPAs are designated areas of the ocean where human activities are regulated to conserve biodiversity, protect habitats, and restore fish stocks.

249

MPAs are established by governments to protect marine life and ecosystems from overexploitation and damage. They are also designed to preserve critical habitats and allow for the replenishment of fish stocks. MPAs are critical for the long-term sustainability of our oceans and play an important role in maintaining marine ecosystem health.

Effective management of MPAs involves careful planning, monitoring, and enforcement of regulations. This requires collaboration between governments, local communities, and other stakeholders. The success of MPAs also depends on public education and awareness, as well as support from the fishing industry and other user groups.

The establishment of MPAs has been proven to be an effective tool for protecting biodiversity and promoting sustainable fisheries. However, the process of selecting and designating MPAs is complex and challenging, as it involves balancing various interests and conflicting views.

1. - Habitat Conservation

The health and resilience of marine ecosystems depend on the protection and restoration of vital habitats such as coral reefs, mangroves, seaweed beds, and wetlands.

We must prioritize conservation efforts to help mitigate the negative impacts of human activities on these habitats. One way to achieve this is through habitat restoration projects, which aim to revive and enhance the natural functions of these vital ecosystems.

By restoring and protecting these habitats, we can help maintain marine life's delicate balance and promote healthy ecosystem functioning. Habitat mapping is another crucial aspect of conservation efforts.

By accurately mapping out the distribution and extent of these habitats, we can better understand their importance and identify areas in need of protection. This information can also be used to assess the effectiveness of conservation efforts and guide future restoration projects.

Additionally, measures to reduce habitat degradation and loss, such as implementing sustainable fishing practises and reducing pollution, are essential for maintaining the health and resilience of these habitats.

Ultimately, protecting and restoring marine habitats not only benefits the plants and animals that call them home but also has a positive impact on human communities. Healthy marine ecosystems can provide valuable resources, such as fisheries and tourism opportunities, and also play a crucial role in regulating the earth's climate.

250

1. - Species Protection

Marine conservation is critical for preserving endangered species. Due to declining populations of marine mammals, sea turtles, sharks, and seabirds, conservation efforts are crucial to prevent further decline and promote recovery. These efforts include monitoring species, protecting habitat, reducing bycatch, and enforcing wildlife protection laws. Without these measures, the survival of these species would be at risk.

Monitoring species is a vital aspect of marine conservation. It allows for population tracking and identifies any declines or increases. This information is crucial for implementing effective conservation strategies.

Protecting habitats is also critical to the survival of endangered marine species. By preserving their habitats, we can ensure that they have a safe and suitable environment to thrive in.

Additionally, reducing catch, which is the unintentional capture of non-target species, can greatly benefit endangered marine species. By minimising catches, we can reduce the impact on their populations and help them recover.

Enforcing wildlife protection laws is another critical aspect of marine conservation. These laws are in place to prevent exploitation and harm to endangered species.

Enforcing these laws can ensure that these species are not targeted for trade or other harmful activities.

Overall, conservation efforts are vital for the protection and recovery of endangered marine species. By implementing measures such as monitoring, habitat protection, catch reduction, and law enforcement, we can make a significant impact on the survival of these species.

251

1. - Marine Pollution Control

Marine pollution is a growing concern for the world's oceans. With various sources contributing to the problem, it is important that we take action to protect our marine ecosystems and biodiversity.

This means implementing measures that will help reduce pollution, improve waste management practises, and promote recycling and circular economy principles. By doing so, we can help preserve the health of our oceans and the many species that rely on them.

One of the main sources of marine pollution is shipping activity. As goods are transported across the world's oceans, they contribute to pollution through the release of harmful chemicals and waste products are released, contributing to pollution.

In addition, industrial discharges and agricultural runoff also play a significant role in polluting our oceans.

These activities release toxins and other pollutants into the water, which can have devastating effects on marine life.

Furthermore, plastic waste is another major contributor to marine pollution, with millions of tonnes of plastic ending up in our oceans each year. Plastic does not break down easily, so it poses a serious threat to marine ecosystems and can cause harm to marine animals through entanglement or ingestion.

To combat marine pollution, we must take steps to reduce our impact on the marine environment. : Stricter regulations on shipping activities, improved waste management practices, and the promotion of recycling and circular economy principles can achieve this.

By reducing the amount of pollution entering our oceans, we can help protect the delicate balance of marine ecosystems and ensure the survival of the many species that call them home.

252

1. - Climate Change Adaptation

Climate change is a global issue that has severe implications for our planet's marine ecosystems and coastal communities. Rising sea levels, ocean acidification, and extreme weather events are only a few of the many consequences of changing climate patterns.

These changes pose significant challenges that need immediate attention. To minimise the impact of climate change, adaptation strategies are crucial and come in various forms. Coastal protection measures are one of the most effective ways to combat climate change. These measures include creating natural barriers, such as sand dunes and mangrove forests, to protect the coast from storm surges and erosion. Additionally, building artificial structures, like seawalls and breakwaters, can also help prevent damage to coastal communities. These adaptation strategies not only protect the coast but also help preserve delicate marine ecosystems.

Ecosystem-based approaches are another critical aspect of climate change adaptation. This involves working with natural ecosystems such as coral reefs and wetlands to help mitigate the impacts of climate change. These ecosystems provide valuable services, such as storm protection, carbon sequestration, and water filtration, that are essential for the well-being of coastal communities. By protecting and restoring these ecosystems, we can increase their resilience to climate change and provide a sustainable solution for adaptation.

253

1. - Sustainable fisheries management

Overfishing, illegal fishing, and destructive fishing practises have become major issues for the marine ecosystem. These activities have resulted in a rapid decline in fish stocks, disrupting the marine food webs and degrading the habitats of several marine species.

This has had a significant impact on the overall health of the ocean, leading to a decrease in fish populations and affecting the resilience of the ecosystem. As a result, it has become crucial to implement sustainable fishing management strategies to address these challenges.

Science-based approaches are one of the key strategies for sustainable fisheries management. By relying on scientific evidence and data, fisheries managers can make informed decisions to regulate fishing activities and ensure the long-term sustainability of fish populations.

Furthermore, promoting sustainable fishing practises such as selective fishing methods and avoiding overfishing can help maintain healthy fish populations. It is also essential to address the issue of illegal, unreported, and unregulated fishing, which often goes unnoticed and can have a severe impact on fish stocks.

By actively combating these activities, we can protect marine habitats and promote sustainable fishing practises. In conclusion, sustainable fishery management is crucial for the health of the ocean and the survival of marine species.

By implementing science-based strategies, promoting sustainable fishing practises, and addressing illegal fishing activities, we can ensure the long-term sustainability of fish populations and support the resilience of the marine ecosystem. It is essential to prioritise these efforts to protect the ocean and the livelihoods of those who depend on it.

254

1. - Maritime Spatial Planning

Maritime spatial planning is a crucial aspect of coastal and marine management. It aims to balance the various activities taking place in these areas while promoting sustainable development and preserving marine biodiversity.

By considering multiple uses of the marine environment, such as shipping, fishing, energy production, conservation, and recreation, it ensures that the needs of all stakeholders are taken into account. One of the key benefits of maritime spatial planning is its ability to minimise conflicts between different marine users.

By carefully mapping out the various activities and their locations, potential conflicts can be identified and mitigated beforehand. This not only helps maintain peace and harmony among stakeholders but also contributes to the sustainable use of marine resources. In addition to promoting sustainable development and minimising conflicts, maritime spatial planning also plays a crucial role in conserving marine biodiversity.

By considering the various uses of the marine environment and their potential impact on the ecosystem, it helps ensure the long-term health and sustainability of marine species and habitats. This is essential for maintaining the delicate balance of marine ecosystems and preserving them for future generations. Overall, maritime spatial planning is an essential tool for achieving sustainable development and the conservation of oceans and coasts.

255

1. - Public Awareness and Education

The environment is a shared resource that affects all of us, and it is our responsibility to take care of it. By raising awareness and promoting environmental education and stewardship among stakeholders, we can ensure that everyone understands the importance of preserving our natural resources.

This includes policymakers, who have the ability to create and implement laws and regulations that can protect our environment. It also includes industry players who can make a big impact through sustainable practises.

Coastal communities, which depend on the ocean for their livelihood, must also be included in these efforts. Of course, the general public, whose actions and choices can greatly impact the environment, must also be educated and engaged. Through these efforts, a culture of conservation and sustainable use of marine resources can be fostered.

This means finding a way to use our ocean's resources without depleting them, so that future generations can also benefit from them. By educating and engaging stakeholders, we create a society that values and respects the environment and works toward its preservation. This not only benefits the marine ecosystem but also the people who depend on it for their livelihoods and well-being.

256

1. - Exclusive Economic Zones (EEZs)

The United Nations Convention on the Law of the Sea (UNCLOS) recognizes Exclusive Economic Zones (EEZs) as maritime territories.

These zones give coastal states the right to control and regulate the use of marine resources within a specific area extending from the shoreline.

This has been a significant development in international law and has helped coastal states to protect their marine resources and manage them in a sustainable manner.

The concept of EEZs was first introduced in the 1982 UNCLOS and has since been adopted by 168 countries. The economic importance of these zones cannot be overstated, as they cover over 38% of the world's oceans and are home to some of the most valuable and diverse marine resources. These resources can include fish, oil, gas, and minerals, among others.

The establishment of EEZs has also helped reduce conflicts between neighbouring countries over the resources found in the oceans. Furthermore, EEZs have also played a crucial role in promoting international cooperation and protecting the marine environment.

Coastal states are responsible for managing and preserving the marine resources within their EEZs, which has led to the implementation of sustainable fishing practises and the protection of endangered species. This has also allowed for the development of scientific research and exploration in these zones, leading to a better understanding of the ocean's ecosystems.

An exclusive economic zone (EEZ) is a legally defined area of the sea that extends to a maximum of 200 nautical miles from a coastal state's baselines. The United Nations Convention on the Law of the

257

Sea (UNCLOS) recognizes this zone and intends to give coastal states control over the resources and activities in the area.

The baseline typically corresponds to the low-water line along the coast, but in some cases it can be extended if the state can prove the presence of a continental shelf.

The primary purpose of the EEZ is to give coastal states the exclusive rights to exploit, conserve, and manage the natural resources in the area, including fish, oil, gas, and minerals. This allows coastal states to benefit economically from their coastal resources and also gives them the authority to regulate activities such as fishing, navigation, and scientific research within their EEZ.

UNCLOS also allows for the extension of an EEZ beyond the standard 200 nautical miles in certain circumstances. This is referred to as the "extended continental shelf " and is based on the presence of a natural prolongation of the coastal state's land territory. This allows coastal states to claim a larger portion of the sea and its resources, but it must be supported by scientific evidence and approved by the Commission on the Limits of the Continental Shelf.

The EEZ is an important tool for coastal states to protect their interests and assert their sovereignty over the resources in their surrounding waters.

Within their EEZs, coastal states have sovereign rights over the exploration, exploitation, conservation, and management of natural resources, including fish stocks, minerals, oil, gas, and renewable energy resources. This means that the coastal states have the power to control and regulate activities within their jurisdiction, such as fishing, mining, and drilling, to ensure the sustainable use and protection of these resources. These resources are crucial for the economic development, food security, and livelihoods of coastal communities.

One of the main purposes of the EEZ is to provide coastal states with exclusive rights to exploit and conserve the living and non-living resources within their waters. This is important because it allows coastal states to have control over their resources and prevent exploitation by foreign vessels or companies.

This also enables coastal states to manage their resources sustainably, ensuring that future generations can also benefit from them.

Furthermore, the EEZ provides coastal states with the power to enforce international laws and regulations to protect their marine resources. This includes measures to prevent overfishing, pollution, and other harmful activities that could damage the marine environment. These measures are essential to maintain the health and resilience of the marine ecosystem, which supports the livelihoods of millions of people around the world. In conclusion, the EEZ is a vital tool for coastal states to manage and protect their natural resources, allowing them to benefit from sustainable use while preserving them.

As the Exclusive Economic Zone (EEZ) is an area extending from the coastal state's territorial sea, it is essential to note that this zone is intended for the coastal state's exclusive economic use. However, the United Nations Convention on the Law of the Sea (UNCLOS) also guarantees the freedom of navigation and overflight for all states, including military vessels and aircraft, through these zones. This means that while the coastal state has exclusive rights over the resources within their

EEZ, other states can still navigate and fly through these zones.

The right of innocent passage is granted to foreign ships and aircraft in the EEZ, as long as they do not pose a threat to the coastal state's security or environment. This means that these ships and aircraft can pass through the EEZ without any interference from the coastal state, as long as they do not engage in any activities that may harm the coastal state's security or environment.

This right of innocent passage is an important aspect of UNCLOS, as it ensures that all states have the freedom to navigate and fly through the EEZ without any restrictions. In conclusion, while the EEZ grants exclusive economic rights to the coastal state, UNCLOS also guarantees the freedom of navigation and overflight for all states.

This means that foreign ships and aircraft can pass through the EEZ without any restrictions, as long as they do not pose a threat to the coastal state's security or environment. This balance between exclusive rights and freedom of navigation is an essential aspect of UNCLOS and is crucial in maintaining peaceful and cooperative relations between coastal and non-coastal states.

Coastal states around the world have the privilege of exercising sovereign rights over the living and non-living resources that lie within their Exclusive Economic Zones (EEZs). This essentially means that these states have complete control over the management and regulation of activities that take place in these areas.

These rights enable them to harness the resources within their EEZs for the benefit of their own citizens, while simultaneously safeguarding these resources from exploitation or damage by external entities.

One of the main reasons for establishing EEZs was to extend the control of coastal states beyond their territorial waters, which usually only extend up to 12 nautical miles from the shore. This expansion of jurisdiction has allowed states to tap into the vast resources that lie within their EEZs, such as fisheries, oil and gas reserves, and even minerals on the seabed.

This has also given them the power to regulate activities like fishing, drilling, and mining in order to maintain the sustainability of these resources for future generations.

Furthermore, the EEZ concept has also helped reduce conflicts between neighbouring states over resource ownership. By clearly defining the boundaries of their EEZs, coastal states can avoid any misunderstandings or disputes with other countries and instead focus on managing and protecting their resources. This has also led to increased cooperation and collaboration between states in the management of shared resources, promoting peaceful and sustainable development in these areas.

Coastal states are responsible for protecting and preserving the marine environment within their Exclusive Economic Zones (EEZs). This is the area that extends up to 200 nautical miles from the country's coastline, where the state has special rights for exploiting, conserving, and managing natural resources.

This means that coastal states must take necessary preventative and control measures to ensure the health of the marine environment and its resources. One of the primary responsibilities of coastal states is to prevent and control marine pollution within their EEZs. This includes taking measures to reduce pollution from land-based sources, such as industrial and agricultural activities, as well as maritime activities, such as shipping and oil exploration. By doing so, coastal states can protect the marine environment and the livelihoods of those who depend on it.

In addition to preventing pollution, coastal states also have a duty to conserve marine biodiversity within their EEZs. This includes protecting and managing the various species and habitats that exist in these waters. By doing so, coastal states can help maintain the health and balance of the marine ecosystem, which is essential for the sustainability of fisheries and other marine resources. The use of the Exclusive Economic Zones (EEZs) is not just limited to resource exploration and exploitation. In fact, it also allows for the laying of submarine cables and pipelines to facilitate

communication and transportation.

While these activities are allowed, they have to be carried out in full compliance with international law. Therefore, any laying of cables or pipelines in these zones must be done with utmost care, ensuring that they do not interfere with the rights of the coastal state. This is crucial as it helps maintain the balance between the interests of the coastal state and those of other countries. Additionally, it guarantees that the exploitation of these zones does not benefit a select few but rather serves the greater good of all

parties involved.

The laying of cables and pipelines in EEZs is an important aspect of international law that is often overlooked. It not only promotes communication and transportation but also helps in the development of these zones.

This expansion of activities in EEZs not only benefits the coastal state but also opens up new opportunities for international cooperation and partnerships. By adhering to international laws and respecting the rights of the coastal state, the laying of cables and pipelines in EEZs can bring about significant progress and growth in the region.

The United Nations Convention on the Law of the Sea (UNCLOS) plays a vital role in preventing conflicts over Exclusive Economic Zones (EEZs). This is done through various mechanisms, such as negotiation, mediation, arbitration, and adjudication. These methods provide the necessary framework for peaceful settlement of disputes that may arise from claims over EEZs.

With the ever-increasing demand for resources, the UNCLOS serves as a crucial tool in upholding international law and ensuring equal access to the world's oceans.

Through the UNCLOS, countries have a platform to present their cases and resolve disputes in a peaceful manner.

Negotiation, the most common method used, allows parties to come to an agreement through direct communication.

Mediation, on the other hand, involves a neutral third party who helps facilitate communication between conflicting parties.

If these methods fail, the UNCLOS also provides for binding decisions through arbitration and adjudication.

These mechanisms ensure that disputes are settled in a fair and impartial manner, thereby maintaining the integrity of the EEZ system. The UNCLOS has played a significant role in reducing the potential for conflicts over EEZs. It promotes cooperation and peaceful resolution of disputes, thereby preventing any escalation of tensions.

1. - Climate changes

Climate change is a global issue that affects not only the land but also the vast expanses of the ocean. The repercussions of climate change on the sea are vast and have far-reaching consequences that are not just limited to the environment but also have a significant impact on the ecological and socio-economic systems.

As the Earth's temperature continues to rise, the ocean is bearing the brunt of it, leading to various changes that are affecting the delicate balance of marine life. The most significant impact of climate change on the ocean is the rise in sea levels. As the polar ice caps continue to melt, the sea levels are rising at an alarming rate, which has already led to the displacement of many coastal communities.

This has also resulted in the loss of critical habitats for various marine species, leading to a decline in their populations. Moreover, the ocean is also becoming more acidic due to the increased absorption of carbon dioxide, which is having a detrimental effect on coral reefs and other marine life. Another important repercussion of climate change on the ocean is the increase in extreme weather events such as hurricanes, typhoons, and cyclones.

These events not only cause massive destruction on land but also have severe consequences for marine life. The strong winds and heavy rains can lead to the destruction of coral reefs and other critical habitats, leaving many marine species vulnerable and at risk of extinction.

As climate change continues to worsen, the ocean will continue to face significant challenges, and it is crucial for us to take immediate action to mitigate its effects.

264

1. - Rising Sea Levels

One of the most pronounced effects of climate change on our oceans is the alarming rise in sea levels. As the Earth's temperatures continue to increase, the melting of glaciers and ice sheets has a direct impact on the expansion of seawater, leading to coastal inundation.

This phenomenon not only causes significant coastal erosion but also poses a severe threat to low-lying areas, coastal communities, and vital infrastructure, not to mention the delicate balance of our marine ecosystems. The consequences of rising sea levels are far-reaching and can wreak havoc on our planet's fragile ecosystems. The inundation of coastal areas not only displaces human populations but also poses a significant risk to the diverse flora and fauna that call these areas home. Furthermore, the destruction of vital infrastructure such as roads, ports, and airports severely hinders economic growth and development.

The impact on coastal communities cannot be overstated, as the livelihoods of millions of people are dependent on these regions. As sea levels continue to rise, the urgency to address climate change becomes all the more pressing. It is clear that the rising sea levels caused by climate change have far-reaching consequences, and the situation demands immediate action.

The devastating effects on both human and marine life are impossible to ignore, and it is crucial to address this issue before it becomes irreversible. Transitioning to sustainable practices that reduce greenhouse gas emissions and protect our oceans and the planet as a whole is imperative.

It is time for individuals, communities, and governments to come together and take meaningful action to mitigate the impacts of climate change on our oceans.

265

1. - Ocean Warming

Climate change is a global phenomenon that has many consequences, one of which is the warming of the ocean. As the Earth's temperature rises, the ocean is also affected and its surface temperatures increase.

This is primarily due to the absorption of heat-trapping greenhouse gases, which are released into the atmosphere through human activities such as burning fossil fuels. The consequences of this warming are far-reaching and have significant impacts on marine ecosystems. The warming of the ocean has a domino effect on marine ecosystems. As temperatures rise, it disrupts the delicate balance of these ecosystems and causes changes in ocean circulation patterns. This, in turn, affects the distribution, migration, and behaviour of marine species. For example, some species may need to move to cooler waters to survive, while others may struggle to adapt to the changing conditions.

This disruption can also lead to a decline in certain species, which can have a cascading effect on the entire ecosystem. Furthermore, the warming of the ocean has other implications that can harm marine biodiversity and ecosystem health. For instance, warmer waters can fuel the intensity and frequency of marine heat waves, which can have devastating effects on marine life.

These heat waves can also lead to coral bleaching events, where reefs lose their vibrant colours and become more susceptible to disease and death. Additionally, warmer waters can also contribute to the growth of harmful algal blooms, which can have toxic effects on marine organisms and their habitats. Overall, the warming of the ocean is a significant consequence of climate change that poses a serious threat to the health and stability of marine ecosystems.

266

1. - Ocean Acidification

Ocean acidification is a serious threat to the world's oceans and the creatures that call it home. The rising levels of carbon dioxide in the atmosphere are not only responsible for global warming but also have a direct impact on the pH levels of seawater.

As CO2 dissolves in the ocean, it lowers the pH and increases the acidity of the water. This change in acidity can have devastating effects on marine organisms, particularly those that rely on calcium carbonate to build and maintain their shells or skeletons. The consequences of ocean acidification extend far beyond just the creatures that inhabit the ocean.

As these organisms struggle to adapt to the changing conditions, it can have a ripple effect throughout the entire marine ecosystem. Fish stocks may decline, biodiversity may be impacted, and the overall functioning of the ecosystem may be compromised. This not only affects the health of the ocean but also has implications for human communities that rely on it for food and livelihoods. It is imperative that we address the issue of ocean acidification in order to protect the delicate balance of our oceans and the life within them.

The impact of ocean acidification is not limited to just the ocean. It also has far-reaching consequences for the Earth's climate and the planet as a whole. As marine organisms struggle to survive in the increasingly acidic waters, the delicate balance of the ocean and its ability to absorb carbon dioxide may be disrupted.

This could exacerbate the effects of global warming, leading to even more severe and catastrophic consequences for our planet. It is crucial that we take action to reduce carbon emissions and mitigate the effects of ocean acidification in order to preserve the health of our oceans and the Earth itself.

267

1. - Changes in Marine Habitats

Climate change is not a singular event but a continuous cycle that alters marine habitats and ecosystems. These changes are caused by shifts in temperature, currents, and nutrient availability, which in turn affect the distribution and composition of marine species. The most vulnerable of these ecosystems are coral reefs, mangroves, sea grass beds, and polar ecosystems.

Due to the impacts of climate change, these areas are facing a multitude of threats, such as coral bleaching, habitat loss, and species decline. The consequences of these changes extend far beyond just the marine species. They have a profound impact on human activities such as fisheries, coastal protection, and the overall ecosystem services provided by marine ecosystems.

As the distribution and composition of marine species change, it directly affects the availability of seafood for fishermen. In addition, the loss of coral reefs and mangroves also means a loss of natural barriers that protect coastlines from storms and erosion. Furthermore, these ecosystems play a crucial role in providing valuable services such as carbon storage and water filtration. It is imperative to address the issue of climate change and its impact on marine habitats and ecosystems.

By understanding the complexities of these changes, we can make informed decisions and implement sustainable solutions to mitigate the effects. It is essential to protect and preserve these vital ecosystems not only for the sake of marine species but for the well-being of our planet and all those who depend on it.

268

1. - Extreme Weather Events

Climate change is a major concern for coastal communities, and it is a problem that is only getting worse. As extreme weather events become more frequent and intense, the risk to these communities also increases. It is not just the coastal infrastructure that is at risk, but also the lives and livelihoods of the people who call these areas home.

This is a truly alarming situation, and it is one that requires urgent action. One of the main risks that coastal communities face is the potential for property damage. With hurricanes, typhoons, and storm surges becoming more frequent and unpredictable, homes and businesses are at risk of being completely destroyed.

This not only leads to the displacement of residents but also disrupts their daily lives. In addition, the loss of life is a major concern, as these extreme weather events can have devastating impacts on human safety. The vulnerability of coastal communities to these impacts highlights the urgent need for action to address and mitigate the effects of climate change.

The effects of climate change are not only limited to the physical damage caused by extreme weather events. These impacts also have far-reaching consequences for the environment and the livelihoods of those living in coastal communities. With rising sea levels, coastal erosion, and the destruction of marine ecosystems, the very foundation of these communities is at risk.

269

1. - Melting Polar Ice Caps

Climate change is a pressing issue that has been consistently impacting the Earth's Polar Regions. The continuous rise in global temperatures has caused the melting of polar ice caps and glaciers at an alarming rate. As a result, the Arctic sea ice and Antarctic ice shelves are retreating, causing a significant change in the Earth's ecosystem.

The changes in the polar ice cover have far-reaching consequences, including the rise of sea levels, alteration of ocean salinity and circulation patterns, and disruption of regional climate systems. The melting of polar ice also has a profound impact on the diverse and fragile ecosystems of the Arctic and Antarctic regions.

As the ice continues to melt, it affects the survival and reproduction of various species of wildlife, causing a decline in their populations. Moreover, indigenous communities that rely on ice-dependent resources for their livelihood are also facing significant challenges due to the changes in polar ice cover.

These communities are experiencing adverse effects on their cultural practises, traditional knowledge, and overall well-being. It is crucial to address the issue of climate change and its impact on polar ice cover before it reaches a point of no return. The rapid melting of polar ice is a clear indication that our planet is in grave danger. Unless we take immediate action to reduce our carbon footprint and make sustainable choices, the consequences of climate change will continue to intensify, causing irreversible damage to our planet and its inhabitants.

270

1. - Impacts on Marine Industries

Climate change is very detrimental to the world's oceans and the marine industries that rely on them. These industries are struggling to adapt to the changing ocean conditions, which are having a severe impact on their operations. Marine industries like fishing, aquaculture, shipping, tourism, and offshore energy production are experiencing the far-reaching effects of global warming.

As the Earth's temperature rises, so does the temperature of the oceans. This has a direct effect on the fish stocks, marine habitats, and maritime operations, which all rely on stable ocean conditions to thrive. The rise in ocean temperatures is causing widespread damage to marine ecosystems. As the water becomes warmer, it becomes less oxygenated, which can lead to the destruction of coral reefs and the depletion of fish populations.

These changes have a knock-on effect on the marine industries that rely on these resources. For example, the fishing industry is experiencing a decline in catches, leading to economic losses for fishermen. Similarly, the tourism industry is being affected as marine habitats deteriorate, causing a decrease in the number of tourists who come to visit these areas.

The offshore energy production industry is also facing challenges as changes in ocean conditions can damage equipment and disrupt operations. As a result, coastal economies are facing significant adaptation costs as they struggle to cope with the impacts of climate change on their marine industries.

271

1. The Navy

The national defence plays a crucial role in protecting the country's security and interests in both domestic and external contexts. As part of this defence, the Armed Forces have been entrusted with one of the most critical responsibilities - ensuring the nation's safety.

The Armed Forces comprises different branches, such as the navy, that work together to achieve this goal. The naval forces, being a part of the Armed Forces, hold a significant responsibility in securing the country's coastal borders. They work closely with other branches to safeguard the country's maritime interests.

The navy is equipped with advanced technologies and specialized personnel to monitor and protect the nation's shores from any potential threats. With advanced surveillance systems and a fleet of ships, the navy ensures that the country remains secure from any external attacks.

The national defence strategy is not limited to protecting the country's boundaries but also extends to safeguarding its interests. The navy has been tasked with protecting the country's maritime resources and ensuring they are not exploited by external threats.

They also work toward maintaining peace and stability in the surrounding regions by conducting joint operations with other countries. The naval forces play a crucial role in the country's national defence and remain ever-vigilant to protect the nation's security and interests.

272

1. - The Military use of the Sea

The military use of the sea, also known as naval or maritime warfare, is a vital tool for any nation that has access to it. The sea is a vast and expansive body of water that has been a part of human history since the dawn of civilisation.

Over time, it has become a key element in military tactics and strategy, with naval forces utilising it to achieve various military objectives. These objectives can range from protecting trade routes and securing vital resources to conducting offensive operations against enemy forces.

One of the key components of naval warfare is the use of naval vessels, such as aircraft carriers, destroyers, and submarines, to project power and control the sea. These vessels are equipped with a wide range of weapons and technologies, such as missiles, torpedoes, and radar systems, making them highly versatile and effective in combat situations. In addition to these vessels, naval forces also utilise aircraft, such as fighter jets and helicopters, to provide air support and reconnaissance capabilities. Naval warfare also involves a complex network of supply and logistics to support naval operations. This includes the maintenance and repair of ships, the transportation of supplies and personnel, and the establishment of bases and facilities in strategic locations.

Additionally, naval forces must also train and prepare their personnel for the rigours of sea life and combat, ensuring that they are ready for any situation that may arise. Overall, the military use of the sea is a multifaceted and dynamic aspect of warfare that continues to evolve and shape the course of history.

273

1. - Power Projection

Naval deployment is a crucial part of a nation's military operations. It involves the deployment of various naval assets, such as ships, submarines, and aircraft, to strategic regions across the world. These deployments serve multiple purposes, including projecting military power and influence, demonstrating presence, and deterring aggression.

One of the main objectives of naval deployment is to showcase a nation's military might and presence in key areas. This serves as a warning to potential aggressors and helps maintain stability in the region. Additionally, naval forces also play a crucial role in supporting diplomatic, humanitarian, and security operations. This includes providing aid and support during natural disasters, conducting peacekeeping missions, and participating in joint military exercises with allied nations.

Naval deployment is an essential aspect of national security, as it helps protect a country's interests and maintain global stability. It requires careful planning and coordination to ensure that the right assets are deployed in the right locations at the right time. Additionally, naval forces must also be prepared to respond to any potential threats or conflicts that may arise.

Overall, naval deployment plays a significant role in maintaining peace and stability around the world, making it a crucial aspect of a nation's military strategy.

274

1. - Control of Sea Lanes

Maritime nations engage in strategic control of vital sea lanes, chokepoints, and maritime chokepoints to protect the sea. The main focus of these nations is to safeguard shipping, commerce, and trade routes. As such, navies are kept busy conducting patrols, surveillance, and escort missions.

This is done to ensure the safe and unimpeded passage of merchant vessels, oil tankers, and cargo ships through strategic waterways such as the Strait of Hormuz, the Suez Canal, and the Malacca Strait. The significance of maritime chokepoints cannot be overstated. They represent crucial arteries of sea traffic that are vital to the global economy. As such, nations with access to these chokepoints have a significant advantage in controlling trade and commerce.

The ability to monitor and regulate these strategic waterways is of utmost importance to maritime nations. To ensure the free flow of goods and services through these vital shipping lanes, navies often deploy to conduct regular patrols and surveillance in the vicinity of these chokepoints.

275

1. - Sea Denial

Naval forces employ sea denial strategies to prevent or disrupt enemy access to maritime areas, deny adversary freedom of movement, and protect territorial waters and exclusive economic zones. This may involve mine warfare, submarine operations, anti-ship missiles, and naval blockades to hinder enemy shipping and logistics.

Sea denial strategies are crucial in maintaining control over important maritime areas and preventing enemy forces from gaining access to them. These strategies can also be used to hinder enemy movements and disrupt their supply lines, making it difficult for them to carry out their military operations effectively. One of the most common sea denial strategies used by naval forces is mine warfare. Mines are strategically placed in maritime areas to create barriers and prevent enemy ships from entering or leaving. These mines can be triggered by contact, sound, or magnetic forces, making them highly effective in hindering enemy movement.

Another important aspect of sea denial strategies is submarine operations. Submarines are capable of staying underwater for extended periods of time and can launch surprise attacks on enemy ships, making them a valuable asset in sea denial operations. In addition to mines and submarines, naval forces also use anti-ship missiles to disrupt enemy shipping and logistics.

These missiles are designed to target and destroy enemy ships, making it difficult for them to transport troops and supplies. Naval blockades are also an effective sea denial strategy, where naval forces prevent enemy ships from entering or leaving a specific area. This hinders enemy movements and can also prevent them from accessing important resources. Overall, sea denial strategies play a crucial role in maintaining control over maritime areas and disrupting enemy operations.

1. - Amphibious Operations

276

277

Amphibious operations are an essential part of naval warfare, allowing navies to project their military force directly onto enemy territory. These operations involve the use of specialised ships such as amphibious assault ships, landing craft, and amphibious vehicles to transport troops, equipment, and supplies from sea to land. The goal of such operations is to quickly establish a foothold on enemy territory, known as a beachhead, and then conduct joint land, sea, and air operations to eliminate hostile targets.

The success of amphibious operations relies heavily on the coordination and cooperation between different branches of the military. This includes the navy, which provides the means of transportation and protection, as well as the army and air force, which provide ground troops and air support.

Together, these forces work toward the common goal of establishing control over enemy territory and expanding their military presence. Conducting amphibious operations requires careful planning and execution, as it involves navigating complex and often hostile littoral environments.

The ability to quickly deploy and sustain ground forces in these environments is crucial to the success of such operations. Therefore, navies must constantly train and adapt their strategies to remain effective in this type of warfare.

1. - Anti-Access/Area Denial (A2/AD)

Navies around the world develop anti-access and area denial capabilities to deter or thwart enemy naval operations. This is done in order to limit adversary power projection and defend maritime territory and interests.

A2/AD systems are a set of military strategies developed to deny access to hostile forces and protect sovereign waters. These systems may include anti-ship missiles, naval mines, coastal defence batteries, and anti-aircraft defences. The use of A2/AD systems is an essential aspect of modern naval warfare. It allows ships to control and protect their own waters while also preventing enemy ships from entering their territory.

This is especially important for countries with large coastlines and maritime interests. A2/AD capabilities can also be used to protect vital shipping routes and trade routes, ensuring the flow of goods and resources. In addition to their defensive capabilities, A2/AD systems also serve as a deterrent against potential adversaries. By having these systems in place, navies can show their strength and ability to protect their territory, deterring any aggressive actions from other countries.

This also allows for a balance of power and prevents any one country from dominating the seas. Overall, A2/AD capabilities play a crucial role in maintaining peace and stability in the maritime domain.

278

1. - Maritime Security Operations

The naval forces' primary responsibility is maritime security, which aims to prevent and counter piracy, terrorism, illegal trafficking, and other threats to international peace and stability.

Naval forces are the first line of defence against these threats and work closely with their partners, such as coast guards and law enforcement agencies, to patrol maritime borders and conduct interdictions. This collaborative effort is crucial in maintaining the safety and security of the world's oceans. Maritime security operations are not limited to a single country or region. They require international cooperation and coordination to effectively combat the various threats at sea.

Navies work closely with their international partners to share information, intelligence, and resources to ensure the success of these operations. This collaboration also strengthens relationships between nations and promotes trust and understanding among them.

The enforcement of maritime laws and regulations is another critical aspect of naval security operations. Navies not only protect their own territorial waters but also enforce laws and regulations in international waters to ensure the safety and freedom of navigation for all vessels.

By conducting patrols and interdictions, navies uphold the rule of law and deter potential criminals from exploiting the maritime domain. This contributes to maintaining a stable and secure maritime environment for all countries and their citizens.

279

1. - Humanitarian Assistance and Disaster Relief (HA/DR)

The role of naval forces goes beyond mere defence and security. In times of need, they are also the first responders to the call for a humanitarian crisis. In the face of natural disasters, it is the navy that takes the lead in providing much-needed support and aid to those affected. In the midst of despair, the naval vessels serve as a beacon of hope and a sign of relief to those who have lost everything.

When disaster strikes, sea travel becomes a challenge, but the navy rises to the occasion and uses its expertise to deliver aid and medical supplies to the affected areas.

They also provide food, water, and evacuation support to those in desperate need. With their advanced technology, they are able to coordinate rescue and search operations, bringing hope to those who are lost in the chaos.

Often, they are the only means of communication and transport for those stranded in disaster-stricken areas. But the navy's role does not end there. In the aftermath of a disaster, they work tirelessly to restore essential services in the affected areas.

From rebuilding infrastructure to providing security and stability, the navy plays a crucial role in helping communities get back on their feet. Their selfless efforts and dedication to serving humanity in times of crisis truly make them unsung heroes of our time.

280

1. The Navy

The Navy is one of the major components of the Armed Forces, along with the Army, Air Force, and Marine Corps. It is primarily responsible for conducting naval operations that support national defences and foreign policies.

The Navy is also responsible for maintaining a strong presence in key areas to protect the country's interests and ensure freedom of navigation.

The Navy's role in national defence goes beyond just protecting the country's maritime borders. It also plays a crucial role in projecting power and supporting military operations, mainly in the ocean.

With its fleet of ships and submarines, the Navy has the ability to strike targets on land, sea, and air, making it a critical asset in times of conflict. In addition, the Navy also provides humanitarian aid and disaster relief in times of crisis, showcasing its versatility and importance in maintaining global stability. With its highly trained personnel, advanced technology, and strategic positioning, the Navy is a vital component of national defence and plays a significant role in protecting the country's security and interests.

Its mission is constantly evolving to adapt to new threats and challenges, making it a crucial element in safeguarding the nation's sovereignty and maintaining peace and stability in the Exclusive Economic Zones and in the hydrographical routes mainly in where provides also health services to populations.

Naval forces can be employed for power projection, projecting military power and influence beyond the country's immediate borders. In addition to demonstrating military strength, naval forces can also be used to conduct joint exercises with allies. This not only improves military cooperation and coordination but also serves as a show of solidarity and support among nations. In a world where

281

alliances and partnerships are crucial for security, naval forces play a vital role in maintaining and strengthening these relationships. Furthermore, naval forces can also support diplomatic initiatives by providing a visible and tangible presence in regions of strategic importance like the Antarctic.

Naval forces prioritize rapid response and logistical support in responding to natural disasters and humanitarian crises.

This is particularly important in a country like Brazil, which is prone to natural disasters such as flooding and hurricanes.

In times of need, the naval forces can quickly mobilise and deploy resources to affected areas, ensuring that relief supplies reach those in need as quickly as possible. In addition to providing essential support domestically, they also play a crucial role in international relief efforts. With the ability to transport relief supplies and evacuate civilians, they are often called upon to assist neighbouring countries that have been affected by disasters.

This not only showcases a country's commitment to being a responsible global citizen but also strengthens its relationships with its neighbouring countries. Furthermore, naval forces also have the capability to provide medical assistance to affected populations. This is especially important during humanitarian crises, where access to medical care can be limited or nonexistent. With their specialised training and equipment, naval forces are able to provide much-needed medical aid to those in need, saving countless lives in the process.

Overall, the involvement of Naval forces in humanitarian assistance like participating in UN peace keeping Forces in Haiti and disaster relief operations is a crucial aspect of the country's defence strategy, highlighting its commitment to protecting and serving its citizens and the global community.

Naval forces play a crucial role in aiding and supporting civilians during natural disasters. When calamities plague a nation, naval fleets are deployed to provide crucial assistance in the form of immediate response and logistical support.

In the aftermath of a natural disaster, the affected areas can become isolated and cut off from the rest of the world. In such scenarios, naval forces can use their expertise in navigation and access to waterways to reach remote and inaccessible locations. This allows them to provide vital assistance to those in need and ensure that relief supplies and medical aid reach the affected population in a timely manner.

With their expertise and resources, naval forces are essential in mitigating the devastating effects of natural disasters and providing much-needed relief to affected populations as a crucial part of a nation's military, serving to safeguard the country's maritime resources and ensuring the safety of vital installations, fisheries, and underwater infrastructure.

These resources are essential for a nation's economic stability and security, making it a prime target for potential threats. These threats can come in various forms, from sabotage and terrorism to hostile actions by both state and non-state actors. In today's modern world, where globalization has made connectivity and trade more important than ever, the role of naval forces has become even more critical.

They not only protect their own nation's resources but also contribute to maintaining peace and stability in international waters. Their presence serves as a deterrent to any potential threats from hostile nations or groups, preventing any conflicts from escalating into full-scale wars.

Additionally, naval forces also play a vital role in conducting humanitarian and disaster relief operations, showcasing their flexibility and versatility in serving the needs of their nation and its citizens. Despite the advancements in technology and equipment, naval forces still face numerous challenges in fulfilling their responsibilities. From dealing with unpredictable weather conditions to facing new and evolving threats, they must constantly adapt and improve their strategies and capabilities to ensure the safety and protection of their nation's maritime resources.

This requires extensive training, coordination, and cooperation between different branches of the military and with international partners, highlighting the crucial role of naval forces in safeguarding the world's oceans.

It contribute to the implementation of the country's national security strategy by providing maritime security, supporting military operations, and advancing national interests in alignment with broader defence objectives.

Naval forces around the world typically deploy a variety of vessels tailored to specific roles and missions. These vessels can vary widely in size, capabilities, and functions. Each type of vessel has its own unique purpose and is equipped with specific features to carry out that mission effectively.

For example, large aircraft carriers are designed to project power across vast distances, while smaller frigates are better suited for anti-submarine warfare.

Different types of naval vessels are also used to support humanitarian efforts. In the wake of natural disasters, such as hurricanes or earthquakes, naval ships are often deployed to provide aid and relief to affected areas.

These massive ships are able to quickly transport personnel, supplies, and equipment to areas in need. They can also serve as a base for medical facilities, allowing for emergency aid to be delivered to those in need. Aircraft carriers are a symbol of national strength and power, with their sheer size and advanced technology serving as a testament to a country's military might.

They are also a source of national pride, with each one representing the ingenuity and determination of a nation to protect its citizens and defend its interests. These warships have played a crucial role in many historical conflicts and continue to be a key component of modern military strategy.

21 - Types of Navy vessels

a. Destroyers

Destroyers are essential assets in any naval fleet, serving as the backbone of a country's maritime defence. These vessels are highly versatile and equipped with advanced radar and missile systems, making them formidable in air, sea, and underwater battles.

Their primary focus is on anti-aircraft, anti-submarine, and anti-surface warfare capabilities, making them extremely valuable in protecting against various threats. In addition to their advanced weaponry, destroyers are also equipped with state-of-the-art communication systems, allowing them to coordinate with other ships and aircraft in the fleet.

They serve as the eyes and ears of the fleet, providing crucial intelligence and surveillance capabilities. Destroyers are also capable of launching and recovering various types of aircraft, further enhancing their versatility and combat effectiveness.

Furthermore, destroyers are often used for diplomatic purposes, serving as a visible presence in international waters and showcasing a country's military power and influence. They can also be used for humanitarian missions, providing aid and support in times of crisis. With their wide range of capabilities and strategic importance, destroyers play a crucial role in maintaining peace and stability in the world's oceans.

a. Frigates

Frigates are versatile naval vessels that are smaller than destroyers but play a vital role in safeguarding the seas. These ships are equipped with a wide range of capabilities that allow them to carry out different missions.

One of their primary tasks is to escort merchant ships, ensuring their safe passage through treacherous waters. Moreover, they are crucial in anti-submarine warfare, using advanced technology to

286

detect and neutralise any potential threats. The compact size of frigates makes them These ships have the capacity to transport medical supplies, food, and clean water, and they may also feature medical facilities to cater to those in need. In times of conflict, naval vessels play a vital role in protecting vital waterways and providing support to ground forces.

In addition to their military and humanitarian roles, naval vessels also have an important role in safeguarding the environment. Many naval forces have a dedicated environmental protection unit that is responsible for monitoring and responding to potential environmental hazards, such as oil spills or illegal fishing activities.

These units work closely with local authorities to enforce environmental regulations and protect marine life.

Overall, naval vessels serve a crucial role in maintaining peace and security, responding to crises, and protecting the environment.

a.   Aircraft carriers and multi purposes vessels

Aircraft carriers are also equipped with advanced radar systems and are able to detect incoming aircraft and missiles. These powerful warships are often accompanied by other vessels such as destroyers, frigates, and cruisers, creating a formidable task force that can defend itself against any threat. In addition to their military capabilities, aircraft carriers also play a critical role in humanitarian and disaster relief efforts.

ideal for operations in coastal waters. This enables them to navigate through narrow channels and shallow waters, which are often inaccessible to larger vessels.

As a result, they are well-suited for carrying out presence and deterrence missions. Their mere presence in an area can act as a significant deterrent, preventing any hostile actions from taking place. This makes them an invaluable asset to have in any naval fleet. In addition to their military capabilities, frigates also play a crucial role in maintaining peace and security in the maritime domain.

They are often employed in humanitarian missions, providing assistance during natural disasters or other crises at sea. Furthermore, they also serve as a symbol of a nation's power and prestige, projecting its influence in the international arena.

With their adaptability and versatility, frigates are indispensable in modern naval warfare and have proven to be a vital asset in protecting the world's oceans.

a. Corvettes

Corvettes are small and fast warships that are designed to be used in coastal areas. They are not as heavily armed as larger ships, but they are perfect for tasks such as patrolling and escorting. Because they are smaller and lighter, they have limited endurance and range compared to larger vessels. However, this does not make them any less effective in their designated tasks.

In fact, corvettes are more cost-effective and efficient for certain missions, making them an important part of any naval fleet. One of the main advantages of Corvettes is their speed. Due to their smaller size, they are able to navigate through shallow waters and tight spaces that larger ships cannot. This makes them perfect for coastal defences, as they can quickly respond to any threats that may arise. Their agility also makes them a valuable asset in escort missions, as they can easily keep up with larger ships while providing protection.

Additionally, their limited endurance and range can be seen as an advantage for certain missions, as they can quickly return to base for refuelling and maintenance. In conclusion, corvettes may be smaller and less capable than larger vessels, but they are an essential part of any naval fleet. Their speed, agility, and cost-effectiveness make them perfect for tasks such as coastal defence, patrol duties, and escort missions. Despite their limitations, they are able to effectively carry out their designated tasks, making them a valuable asset in any military operation.

a. Submarines

Submarines have been around for centuries, with the first military submarine being invented in 1620 by Dutch inventor Cornelis Drebbel. However, it wasn't until the 19th century that submarines were used in warfare. These underwater vessels have come a long way since then, with advancements in technology and engineering making them capable of prolonged submerged operations.

They are now classified into various types, each with their own specific purpose and capabilities. One of the most well-known types of submarines is the ballistic missile submarine (SSBN), also known as "boomers." These submarines are armed with nuclear missiles and are designed for long-range strategic missions. They serve as a deterrent against enemy nations and are critical in maintaining global security Conversely, attack submarines (SSNs) excel in anti-submarine and anti-surface warfare, rendering them indispensable for safeguarding naval forces and executing offensive operations.

Finally, diesel-electric submarines (SSKs) are smaller and more agile, making them ideal for coastal defence and intelligence gathering missions. Despite their different classifications and purposes, all submarines share the same defining feature—their ability to operate underwater for extended periods. This allows them to move undetected and carry out missions with precision and stealth.

f) Amphibious tracked cars are armoured vehicles that are built to travel on both land and water. These unique vehicles are used in various military operations, such as transporting troops and equipment to hard-to-reach areas. They have been a vital asset to ground forces, providing them with mobility and protection in harsh terrain and challenging conditions. These versatile machines are equipped with powerful engines and are able to navigate through various types of terrain, including mud, sand, and rocky surfaces.

They have a unique design that allows them to float on water, making them an essential tool for amphibious operations. These vehicles are heavily armoured, providing protection to the crew and passengers from enemy fire and explosives. With their impressive capabilities, amphibious tracked cars have played a significant role in military history. They assist ground troops, giving them a strategic advantage.

These vehicles have also been used in humanitarian missions, delivering supplies and aid to areas affected by natural disasters. Their versatility and durability make them an essential asset in any military operation.

These are essential for the rapid deployment of troops and equipment onto shore, making them a critical asset for the "Marines".

a. Patrol vessels

Patrol vessels are essential in monitoring and protecting coastal waters. With their small and fast design, they are able to swiftly navigate through the ocean, making them perfect for coastal patrol and law enforcement missions. These vessels are also crucial in carrying out counter-piracy operations, ensuring the safety of maritime trade and fisheries.

Their primary purpose is to maintain security and safety in the waters. Patrol vessels are equipped with advanced surveillance technology, allowing them to monitor and detect any suspicious activities in their designated areas.

This makes them a valuable asset in border protection, as they are able to intercept and stop illegal activities such as smuggling and trafficking. In addition to their security roles, patrol vessels also play a crucial role in search and rescue operations. Due to their speed and manoeuvrability, they are often the first responders to emergencies at sea. These vessels are equipped with medical supplies and trained personnel, making them capable of providing immediate assistance to those in need. In summary, patrol vessels are versatile and vital assets in maintaining order and safety in the coastal waters.

a. Mine's countermeasure vessels, also known as MCMVs, are essential in the military and naval industries. Advanced technology and machinery, such as state-of-the-art sonar systems, minesweeping equipment, and remotely operated vehicles (ROVs), equip these highly specialized ships.

Their primary purpose is to detect and neutralise sea mines and other underwater explosives, making them crucial assets in protecting and securing naval operations. The sonar systems on MCMVs are designed to emit sound waves that travel through the water and bounce off objects, allowing for the detection of potential threats such as mines. Trained personnel on board the vessel analyze and interpret this information to determine the size, shape, and location of the object.

The minesweeping equipment, on the other hand, is used to physically remove or detonate the mines, ensuring the safety of the ship and its crew. In addition to their primary function, MCMVs also play a vital role in reconnaissance and surveillance missions.

Their advanced technology allows them to gather and analyse data on underwater conditions, providing crucial information for naval operations. Moreover, their remotely operated vehicles (ROVs) enable them to explore and investigate the ocean floor, making them invaluable in search and rescue missions.

a. Sailboats

Some navies maintain sailboats for recreational sailing and ceremonial events, such as regattas, parades, and naval traditions. In many cases, this is to showcase the rich history and traditions of sailing within the navy.

Sailboats are often used in special events, such as fleet week celebrations, where they can be seen parading up and down the coastline for thousands to see. These ceremonial events are important to the navy, as they allow the public to experience naval traditions firsthand. It also allows for sailors to showcase their skills and knowledge of sailing.

The use of sailboats in these events is also a way for the navy to connect with the public and build positive relationships with the community. This can be especially important for naval forces that are stationed in coastal towns or cities. Sailboats are also used for recreational purposes within the navy.

This allows sailors to unwind and relax after long deployments or intense training exercises. It also provides an opportunity for team building and camaraderie among sailors.

The navy recognises the importance of maintaining a healthy work-life balance for its members, and the use of sailboats for recreational purposes is just one way they achieve this. Overall, the use of sailboats for both ceremonial and recreational purposes plays an important role in the navy's culture and traditions.

a. Hydrographical and oceanographic surveys

The vessels used for hydrographical and oceanographic surveys play a crucial role in understanding the world beneath the surface of the ocean. Scientists can collect data on the physical characteristics of the ocean and seabed using these specialized ships equipped with state-of-the-art instruments and sensors.

With this data, researchers can map the seafloor, study ocean currents, and monitor marine environments, providing valuable insights into the health and dynamics of the ocean.

Hydrographical and oceanographic surveys are essential for a variety of maritime activities, including navigation, resource exploration, and environmental management. Accurate mapping of the seafloor is critical for safe navigation, especially in areas with shallow waters or potential hazards.

Additionally, these surveys aid in the search for valuable resources, such as oil and gas deposits, minerals, and other materials that can be extracted from the ocean floor. Furthermore, understanding the physical characteristics of the ocean and its currents can help with environmental management, allowing for more effective protection and conservation of marine ecosystems.

Overall, vessels used for hydrographic and oceanographic surveys are crucial tools in our efforts to better understand and utilise the ocean. With their advanced equipment and skilled crew, these ships provide valuable data that helps us make informed decisions about our interactions with the marine environment. From navigation to resource exploration to environmental protection, these surveys have a wide range of applications that benefit society as a whole.

- Hydrographical survey ships

Hydrographical survey ships are important vessels in modern-day marine activities. These ships, purpose-built watercrafts, conduct detailed mapping of the seabed and nearshore areas.

Sophisticated equipment like multiband and single-beam echo sounders, side scan sonar, and sub-bottom profilers enable this process. These tools allow for the collection of high-resolution bathymetric data, seafloor imagery, and sediment characteristics, providing a comprehensive understanding of the underwater terrain. In addition to their advanced equipment, hydrographical survey ships also have onboard processing facilities, laboratories, and data analysis software. This allows for real-time analysis and interpretation of the collected survey data. The immediate processing of data is crucial for efficient decision-making and planning in various maritime operations, such as offshore construction, port and harbour developments, and resource

exploration.

Furthermore, the data collected by these ships is also used in the creation of accurate and up-to-date nautical charts, aiding safe navigation for vessels of all sizes. Overall, hydrographical survey ships play a crucial role in the development and management of coastal and oceanic areas.

Their advanced technology and capabilities allow for the collection and analysis of vital information, making them an essential tool for marine research, resource management, and safe navigation. These vessels continue to play a significant role in shaping our understanding of the underwater world and its impact on our daily lives.

Brazil in the river Amazons, known for changing its bed accordingly weather conditions, has the "Sirius" for hydrographical survey in the extended hydrographical rivers and inside waters.

- Oceanographic Research vessels

Oceanographic research vessels are versatile ships capable of conducting a wide range of scientific studies in oceanography, marine biology, geology, and climate science. These vessels are equipped with a suite of oceanographic instruments, including CTD (Conductivity, Temperature, Depth) profilers, plankton nets, sediment corers, and water sampling systems, to collect data on ocean properties such as temperature, salinity, currents, and nutrient levels.

Brazil has for this purpose the following vessels: - H44 – "Ary Rongel"; H40 – "Antares"; "Almirante Graça Aranha"; H35 – "Amorim do Valle"; H36 – "Taurus"; H37 – "Garnier Sampaio"; H38 – "Cruzeiro do Sul"; H39 – "Vital de Oliveira"; H11 – "Aspirante Moura".

K - Hospital vessels

These ships serve as the only means of medical attention for many people, especially in the aftermath of natural disasters or military conflicts These vessels, equipped with highly specialized medical equipment and trained personnel, serve as floating hospitals for those in need.

They are capable of providing essential healthcare services, delivering emergency medical aid, and performing life-saving surgeries.

The significance of hospital vessels cannot be overstated, as they have the potential to save countless lives in situations where traditional healthcare facilities are unavailable. In addition to their critical role in providing immediate medical care, hospital vessels also serve as an essential tool for disaster relief efforts. These ships are equipped to navigate through coastal areas and reach remote locations that are often difficult to access during times of crisis.

They are equipped with all the necessary facilities to accommodate large numbers of patients and carry out medical procedures. In many cases, hospital vessels also serve as a means of transporting patients to safer areas for further treatment. This greatly improves the chances of survival for victims of natural disasters and conflicts, who would otherwise have little access to medical attention.

The role of hospital vessels in humanitarian crises is invaluable. These ships serve as a beacon of hope for communities in need, providing essential medical care and alleviating the suffering of those affected by disasters and conflicts.

Their presence can make all the difference for those who would otherwise have nowhere to turn. With their advanced medical capabilities, hospital vessels offer a lifeline for people in remote and coastal areas, bringing with them the promise of a better tomorrow.

In a vast and uncharted sea, hospital vessels have become a beacon of hope and life-saving services. These ships are equipped with state-of-the-art medical facilities that are capable of providing comprehensive medical services. They are like floating cities, where patients from the most remote corners of the world can receive medical care and treatment.

On board the hospital vessels, patients have access to a range of medical services, from basic primary care to advanced surgical procedures. The ships are equipped with operating rooms, intensive care units, diagnostic laboratories, radiology facilities, and patient wards. Highly trained medical professionals staff these facilities tirelessly to provide the best care possible.The ships are also equipped with the latest medical technology, ensuring that patients receive the most advanced treatment available.

The hospital vessels serve as a lifeline for those in need, providing not just medical care but also hope and comfort. They travel to some of the most remote and underserved areas, providing much-needed services to communities that would otherwise have no access to healthcare. These ships are a testament to the dedication and compassion of the medical community, working tirelessly to make a difference in the world.

Staffed with a dedicated team of medical professionals, they are made up of doctors, surgeons, nurses, pharmacists, laboratory technicians, and support staff who are all highly trained and experienced in providing emergency medical care. They are also skilled in treating trauma and performing surgical interventions in even the most challenging environments. The medical teams on these hospital vessels are specially trained to handle the unique challenges that come with providing medical care on a ship.

They understand the importance of being able to adapt and improvise in difficult situations, as well as being able to work closely together as a team. With the unpredictable nature of being at sea, these medical professionals must be ready for anything and be able to think quickly on their feet to ensure the best possible outcomes for their patients.

Despite the challenging environments they work in, the medical teams on these hospital vessels are dedicated to providing the highest quality of care to those in need. They are passionate about their work and are constantly striving to expand their knowledge and skills in order to better serve their patients.

These teams are truly remarkable in their ability to work under pressure and make a positive impact on the lives of those they treat, even in the most challenging of circumstances.

They are the first responders to provide aid to those affected by natural disasters, humanitarian crises, and conflicts. These medical teams are trained to react quickly and efficiently to ensure that survivors receive the medical care they need in a timely manner.

Their primary goal is to alleviate suffering and save lives during times of crisis. These teams, equipped with specialized medical supplies and equipment, stand ready to address a variety of injuries, illnesses, and medical emergencies.

Whether it's a result of an earthquake, tsunami, hurricane, or armed conflict, these medical teams have the necessary tools and skills to provide immediate assistance to those in need.

They work tirelessly in challenging environments to provide aid and comfort to survivors, often risking their own safety in the process. Their dedication and selflessness are commendable, as they leave their homes and families to travel to disaster-affected areas.

These medical teams provide a glimmer of hope to those who have lost everything in the wake of a disaster. With their quick response and expertise, they play a crucial role in helping communities recover and rebuild in the aftermath of a crisis. Thanks to their bravery and unwavering commitment, countless lives have been saved.

Hospital vessels provide essential medical services to areas that may not have access to quality healthcare. These vessels are equipped to handle a wide range of medical needs, from small medical boats or barges to large hospital ships that can accommodate hundreds of patients and medical personnel.

Depending on the specific missions and needs, hospital vessels vary greatly in size and capacity.

This means that hospital vessels can be tailored to meet the specific needs of the communities they serve, whether it be providing emergency medical care or long-term medical services. This adaptability makes hospital vessels a crucial resource in times of crisis and disaster.

Furthermore, hospital vessels also offer a unique opportunity for medical personnel to gain valuable experience and training. As these vessels serve different regions and communities, medical professionals are exposed to a variety of medical conditions and situations, allowing them to expand their skills and knowledge.

This not only benefits the medical personnel but also the communities they serve, as they receive top-quality medical care from experienced professionals. In conclusion, hospital vessels play a vital role in providing medical services to underserved areas and serve as a valuable training ground for medical professionals.

Logistics hubs are the backbone of healthcare supply chains, providing a crucial link between manufacturers and affected populations. These hubs allow for the proper storage and distribution of essential medical supplies, pharmaceuticals, and humanitarian aid.

Without them, the timely delivery of these lifesaving resources to those in need would be nearly impossible. In times of crisis, such as natural disasters or global pandemics, logistics hubs play a critical role in ensuring the efficient and effective flow of medical supplies. They serve as central points for inventory management, ensuring that the right products are available when and where they are needed most. Additionally, these hubs often have specialised storage facilities, such as temperature-controlled environments, to ensure the integrity and safety of delicate medical supplies. Furthermore, logistics hubs also serve as coordination centres for the transportation of medical supplies and aid.

They work closely with transportation providers, such as airlines and ground transportation companies, to ensure that supplies are delivered to affected areas in a timely and cost-effective manner. This seamless coordination is essential for reaching remote and hard-to-reach areas, where the need for medical supplies may be greatest. Overall, logistics hubs play a vital role in providing essential support to affected populations during times of crisis.

A school vessel is not only a platform for training and education but also for building character and teamwork skills. It is a unique opportunity for students to learn the importance of discipline, hard work, and adaptability in a real-world setting.

These ships are equipped with state-of-the-art facilities and technology, providing students with hands-on experience in every aspect of seamanship, navigation, marine engineering, and maritime operations. This comprehensive training prepares students for a variety of careers in the maritime industry, including merchant shipping, naval service, offshore industries, and maritime logistics.

Moreover, these school vessels also offer a diverse and multicultural environment, allowing students to learn from their peers and instructors from different backgrounds. This exposure to different cultures and perspectives helps students develop a global mindset and prepares them to work in an increasingly international industry.

Students also learn the importance of communication and teamwork as they work together to navigate the ship and complete various tasks. Overall, a school vessel provides a unique and invaluable experience for students pursuing a career in the maritime industry. It not only equips them with the necessary technical skills but also instils important values and qualities that will benefit them in their future careers.

As the saying goes, "smooth seas do not make skilful sailors," and these school vessels provide the perfect environment for students to develop the skills and resilience needed to succeed in the challenging and ever-changing maritime industry.

School vessels are crucial in training future seafarers and enhancing the skills of maritime professionals. The offered programs aim to equip individuals with a comprehensive understanding of the sea and its technicalities.

Basic seamanship training is an essential aspect of the curriculum, which focuses on understanding the basic principles of sailing, water safety, and other seamanship practises.

This training provides the necessary knowledge to operate a vessel and develop fundamental skills essential for a sustainable marine career. Navigation and chartwork constitute another critical aspect of the training programmes offered by school vessels. These courses help students understand how to use maps, nautical charts, and other navigational tools to chart their way through the sea.

They also learn how to identify hazards and plan safe routes while on a voyage. Marine engineering, on the other hand, is an integral part of the curriculum for students who aspire to become engineer's onboard vessels. They learn the functioning of different marine systems and how to troubleshoot them in case of any issues. The training programmes also cover other crucial aspects that are necessary for a successful maritime career.

Safety and survival skills are vital for any seafarer and are taught through various simulations and practical training sessions. This training provides students with the necessary skills to handle emergencies at sea. Moreover, students are also taught about maritime law and regulations to ensure that they have a thorough understanding of the legal framework governing the seas.

School vessels provide students with practical, hands-on experience in operating and maintaining a ship at sea. : Students apply the theories and knowledge they have learned in the classroom to real-life situations as they participate in daily routines and watchkeeping duties.

This allows them to gain a deeper understanding of the concepts being taught while honing their skills in various aspects of ship operations.

Under the supervision of experienced instructors and crew members, students are able to gain valuable hands-on experience in various deck operations. From handling mooring lines to setting up gangways, students are exposed to different tasks that are essential in ship operations.

Additionally, they are also given the opportunity to take part in navigation exercises, where they apply their knowledge in identifying navigational aids, plotting courses, and working with different navigational equipment. Moreover, students also get to experience working in the engine room, where they are exposed to the inner workings of a ship's engine and its various components.

This allows them to understand the complexities of marine engineering and gain practical skills in maintaining and troubleshooting engine problems. Safety drills are also regularly conducted, ensuring that students are well-equipped with the necessary knowledge and skills to handle emergency situations at sea. When students attend a school vessel, they will find an environment that prepares them for real-world scenarios by providing simulated exercises. These exercises include man-overboard drills, firefighting training, abandon ship

demonstrations, and collision avoidance manoeuvres.

By participating in these exercises, students develop critical thinking skills, learn the importance of teamwork, and refine their decision-making abilities in challenging maritime environments. The simulated exercises are a key component of a school vessel's curriculum. The training students receive while on board is essential to their development as future seafarers.

The man overboard drills, for instance, teach students how to react quickly and effectively in emergency situations. By working together to put out a fire, students learn the value of teamwork and the need for cooperation in high-stress situations. The abandonship drills and collision avoidance manoeuvres also help students understand the gravity of these situations and how to handle them with confidence and competence. Overall, these simulated exercises provide students with practical experience that will serve them well in their future careers at sea.

In addition to providing facilities and amenities for students and staff during training voyages, school vessels also act as a second home for the students.

The classrooms and lecture halls are not only a place for learning but also a place for students to build a sense of community with their classmates and instructors. The computer labs and training simulators are where students develop practical skills and gain hands-on experience in various maritime operations.

While the mess halls and sleeping quarters are where students take breaks from their rigorous training, the recreational areas allow them to unwind and bond with their peers. These facilities not only provide a comfortable living environment but also foster teamwork and camaraderie among the students.

Additionally, school vessels may also have medical facilities and laboratories for conducting research and practical training exercises. This provides students with a unique opportunity to apply their knowledge in a real-life setting and prepare them for their future careers in the maritime industry.

School vessels are not just a place for academic learning but also a platform for students to develop important life skills. They learn to live and work in close quarters, manage their time effectively, and adapt to constantly changing environments.

School vessels often undertake training voyages to various ports and regions, allowing students to gain real-world exposure to diverse maritime environments. These voyages also provide students with the chance to test their skills and knowledge by interacting with other navies and learning about diverse shipping practises.

This experience provides a more complete picture of the maritime industry, giving students a deeper understanding of the complexities and challenges involved. Moreover, these training voyages allow students to immerse themselves in different cultures, further expanding their horizons and broadening their perspectives. By interacting with people from different backgrounds and nationalities, students gain a better understanding of the global nature of the shipping industry.

This exposure also helps them develop valuable interpersonal skills as they learn to communicate and collaborate with individuals from diverse backgrounds. In addition, these training voyages often involve simulated scenarios that simulate real-life situations, giving students a taste of what it's like to work on a ship in the open sea. This hands-on experience is invaluable as it allows students to apply their theoretical knowledge in a practical setting.

It also helps them develop critical thinking skills as they navigate through challenges and solve problems in a dynamic and constantly changing environment.

22 - Navy Academies

Navy academies, also referred to as naval academy's or maritime academies, are invaluable institutions that provide individuals with the necessary skills and expertise to excel in various naval roles.

These institutes offer a wide range of academic programmes, hands-on training, and professional development opportunities that are essential for students to thrive in the navy, merchant marine, maritime industry, and other related sectors. The rigorous and specialised training provided by navy academies prepares students for the challenges of naval service.

Through a combination of academic coursework, practical training, and leadership development, students are equipped with the skills and knowledge to excel in various naval roles. From navigating the seas to managing complex operations, the comprehensive education offered by these academies prepares students for a variety of career paths in the navy and maritime industry. Moreover, navy academies foster a sense of discipline, integrity, and service among their students.

These institutions are renowned for their strong emphasis on character development and ethical conduct, instilling in students a deep sense of duty and responsibility toward their nation and fellow citizens. By providing students with a holistic education, navy academies not only prepare them for successful careers but also for a life of service, leadership, and impact.

Navy academies play an important role in shaping the future of military service members. These institutions are known for instilling discipline, leadership, and teamwork in their students.

The training provided at these academies is rigorous, with a focus on physical fitness, drill exercises, and military instruction. This ensures that students are well-prepared for the challenges of military life. At navy academies, students not only learn the necessary skills for military service but also develop important personal traits. The emphasis on discipline helps students become more organised, focused, and responsible. Leadership training teaches students how to effectively lead others and make important decisions.

Teamwork is also highly valued at these academies, as military service often involves working closely with others. By the time student's graduate from these institutions, they are well-rounded individuals who possess both the physical and mental capabilities to serve their country. Navy academies provide training that is not only physically demanding but also mentally challenging. Students are required to push themselves to their limits and constantly strive for improvement. This helps them build resilience and develop a never-give-up attitude, which are crucial in military service. Through this intense training, students also learn the importance of perseverance and determination, which are essential in the face of adversity.

Military academies are prestigious institutions that offer a diverse range of academic programmes and courses. These programmes cover a wide variety of subjects, including naval science, maritime studies, marine engineering, navigation, oceanography, maritime law, and maritime security. By providing such a comprehensive curriculum, the academy aims to produce well-rounded individuals who are equipped with the knowledge and skills necessary for a successful career in the navy.

In addition to traditional classroom instruction, military academy students also have the opportunity to gain practical experience through laboratory work and hands-on training. This combination of theoretical and practical learning allows students to develop a deep understanding of maritime operations and related disciplines.

They are able to apply what they have learnt in the classroom to real-world scenarios, preparing them for the challenges they may face in their future careers.

Moreover, military academy offers a unique learning environment that fosters discipline, leadership, and teamwork. The military academy expects students to adhere to strict codes of conduct and follow a rigorous training program.

This not only prepares them for the career in the navy but also instils important values and qualities that are essential for success in any profession. Overall, a military academy provides a well-rounded and immersive educational experience that prepares students for a rewarding and fulfilling career at sea.

Navy academies are more than just a place to acquire education and technical skills required for the navy. They also aim to develop leadership abilities in their students. The focus is on honing skills and qualities that are essential for effective leadership. These skills are not only useful in the military but also highly sought after in various industries.

Navy academies achieve this goal through a variety of programmes and activities. Some of these include leadership seminars where students are exposed to different leadership styles and techniques. They also offer officer training courses that provide practical knowledge and hands-on experience in leading a team.

Additionally, students have the opportunity to participate in mentoring programmes where they can learn from experienced leaders. Moreover, students are given chances to take on leadership positions within the academy itself. This allows them to apply their knowledge and skills in a real-world setting. These leadership positions may range from being in charge of a small team to leading a larger group of students. This hands-on experience is invaluable for their growth and development as leaders. With such comprehensive leadership development programmes, it is no wonder that navy academies produce some of the most skilled and successful leaders in the military.

Some Navy academies offer graduate education programmes to students who have successfully completed their undergraduate studies. These programs offer advanced training and specialization in specific areas of military studies or naval science.

Students who enrol in these programmes will have the opportunity to continue their academic journeys and further their education in fields such as engineering, leadership, and strategic planning.

These programs also offer master's degrees and postgraduate certificates, giving students the chance to enhance their knowledge and skills in a particular area of interest. In addition to the traditional classroom setting, these graduate education programmes also offer hands-on training and real-world experiences. This allows students to gain practical skills and apply their knowledge in real-life situations. For example, students may have the chance to participate in simulations or training exercises, giving them a taste of what it's like to serve in the navy. This not only prepares them for their future careers but also helps them develop critical thinking and problem-solving skills that are essential in the military.

These programs, designed for students dedicated and driven to excel in their military careers, are highly competitive and selective.

Graduates of these programmes often go on to become leaders and innovators in the Navy, making significant contributions to the defence and security of their countries. By offering graduate education programmes, navy academies are not only providing advanced training to their students but also shaping the future of the military and preparing the next generation of leaders.

CONCLUSION

The sea is a vital resource that provides us with food, transportation, and leisure. However, many people fail to recognize its value and often engage in activities that harm the delicate balance of the ocean's ecosystem.

The lack of awareness and education about the importance of the sea has led to its degradation and destruction. From overfishing to pollution, human activities have caused irreparable damage to the ocean.

To preserve the sea and its resources for future generations, it is crucial to instil ethical and moral values in society. This includes educating people about the impact of their actions on the ocean and promoting responsible behaviour toward it. It is not just about following laws and regulations but also about developing a deep sense of respect and responsibility toward the sea. By doing so, we can ensure sustainable use of its resources and protect the diversity of marine life. Implementing ethical and moral values related to the sea requires a collective effort from individuals, communities, and governments.

It is not something that can be achieved overnight but requires continuous education and awareness campaigns. By working together, we can create a society that values and respects the sea, leading to a healthier and more sustainable ocean. Let us not forget that the sea is not just a source of wealth but also a source of life, and it is our duty to protect it.

Civic education is essential to promote responsible behaviour and spread awareness against pollution in the sea, especially due to plastic waste. It aims to educate individuals on the detrimental effects of plastic waste on marine life and the environment.

One of the key strategies for incorporating civic education is to organise awareness campaigns and workshops to educate the public, especially the younger generation, on the importance of proper waste management and the impact of plastic waste on marine ecosystems.

Another effective way to promote civic education is by partnering with local communities and organisations to organise clean-up drives and recycling programmes. This not only helps to clean up the beaches and coastal areas but also encourages individuals to take responsibility for their actions and actively participate in preserving the marine environment.

It also provides an opportunity to educate the public on the proper disposal of plastic waste and the benefits of recycling. Moreover, incorporating civic education in school curricula can also play a significant role in preventing the pollution of the sea with plastic waste.

By including lessons on environmental conservation and waste management, students can develop a sense of responsibility and become advocates for a cleaner and healthier marine environment. This will not only benefit the sea but also create a more environmentally conscious and responsible society.

Environmental education is a vital component to any school curriculum, as it provides students with the knowledge and understanding of how their actions can have a profound impact on the environment.

By integrating environmental education into school curricula at all levels, students can learn about the impacts of plastic pollution on marine ecosystems and the importance of conservation. This type of education would allow students to gain a deeper understanding of the interconnectedness of the marine food web and how plastic waste disrupts this delicate balance. In addition to learning about the negative effects of plastic pollution on marine life, students would also be exposed to various topics related to plastic degradation processes.

This could include discussions on the different types of plastic and their decomposition rates, as well as the various methods of recycling and reducing plastic waste. By educating students about these processes, they can gain a better understanding of the long-term effects of plastic pollution and the importance of responsible waste management. Furthermore, incorporating environmental education into school curricula would also highlight the role of individuals in reducing plastic waste and promoting conservation efforts.

By providing students with the necessary knowledge and skills, they can become agents of change in their communities and work toward a cleaner and more sustainable future. This type of education would empower students to make environmentally conscious choices and encourage others to do the same, ultimately leading to a significant reduction in plastic pollution and a healthier marine ecosystem.

The ocean can be a fun and relaxing place to take a dip, but it can also be quite dangerous. Without proper precautions, swimming in the sea can pose significant risks to individual's safety. The good news is that there are many civic education efforts aimed at promoting safe swimming practises. These efforts can help raise awareness about the importance of water safety and reduce the incidence of accidents and drowning.

Local government organizations actually run many of these civic education programs, making them accessible and effective.

One of the key ways that civic education programmes promote safe swimming practises is by providing information about the potential dangers of swimming in the sea.

By educating individuals about these risks and providing helpful tips, civic education efforts can greatly reduce the incidence of accidents and drowning in the sea.

The issue of ballast water pollution and waste discharges from commercial vessels is a growing concern that requires a holistic approach to combat. This complex problem demands a well-rounded solution that involves not only technological advancements but also international cooperation, regulatory compliance, and stakeholder engagement.

One of the key factors in addressing this issue is through international cooperation. As commercial vessels travel across various waters, it is vital for countries to work together in implementing regulations and policies to prevent pollution and waste discharges. This involves not only the cooperation of governments but also the collaboration of different industries and organisations to effectively address the problem at a global scale.

Technological innovation is another crucial aspect in tackling this issue. With the advancement of technology, there are now more efficient and environmentally friendly ways to manage ballast water and waste on commercial vessels. This includes the development of new filtration systems and waste management techniques that can reduce the impact of pollution and waste discharges on marine ecosystems.

The consequences for breaking maritime pollution laws can be severe, whether the offender is an individual or a company. Strong law enforcement is crucial in preventing repeated infringements. Without strict penalties, there is little incentive for individuals or companies to comply with these laws. In addition to monetary fines and other legal repercussions, there are more significant consequences for polluting the ocean.

The environment suffers greatly from the irresponsible actions of those who choose to ignore maritime legislation. Marine life is put at risk, and the balance of the ecosystem is disrupted. The livelihoods of those who rely on the ocean for their income are also affected.

These are all important factors to consider when enforcing penalties for maritime pollution. It is not enough to simply have laws in place; they must be effectively enforced to prevent further harm to the environment and those who depend on it. This is where a strong law enforcement system comes into play.

By consistently enforcing penalties and holding individuals and companies accountable for their actions, the repetition of infringements can be stopped. It is crucial that these laws are taken seriously and that consequences are given to those who choose to disregard them.

Bibliography

Alderton, P., 2004. Reeds sea transport: operation and economics. 5th ed. London: Adlard Coles Nautical.

Alderton, P.M., and Winchester, N. conducted a study on port management and operations in 2007.3rd ed. London: LLP.

Anderson, P., 2005. Cracking the code: The relevance of the ISM code and its impact on shipping practises. London: The Nautical Institute.

Anderson, P., 2005. Cracking the code: The relevance of the ISM code and its impact on shipping practises. London: The Nautical Institute.

Barnett, M.L., Gatfield, D., and Pekcan, C.H., 2006. Barriers to progress in maritime education and training. London: The Nautical Institute.

Bellard, C., Bertelsmeier, C., Leadley, P., et al., 2012. Impacts of climate change on the future of biodiversity. Oxford: Oxford University Press.

Bertram, V., 2000. Practical ship hydrodynamics. Oxford: Butterworth-Heinemann.

Black, J., 2009. Naval power: A history of warfare and the sea from 1500 onwards. London: Palgrave Macmillan.

Carlisle, R.P., 2009. The complete guide to careers in special operations. New York: Facts On File.

Charlier, R.H. and Finkl, C.W., 2009. Ocean energy: Tide and tidal power. Berlin: Springer.

Churchill, R., and Lowe, A.V., 1999. The law of the sea. 3rd ed.

Manchester: Manchester University Press.

Couper, A., 1999. Voyages of abuse: seafarers, human rights, and international shipping. London: Pluto Press.

Davis, A.L., 2020. The unseen history of the Navy SEALs: The maritime special operations forces. London: Penguin.

313

Doney, S.C., 2012. Climate change impacts on ocean acidification and marine ecosystems. New York: Springer.

Ehlers, P. and Wolfrum, R., 2008. Ocean governance: A critical assessment. Berlin: Springer.

Ghosh, S., and Bowles, M., 2012. Employment, careers, and innovation in maritime education and training. London: Taylor & Francis.

Hattendorf, J.B., 2013. Maritime strategy and the balance of power: Britain and America in the twentieth century. London: Macmillan.

Hoegh-Guldberg, O., and Bruno, J.F., 2010. The impact of climate change on the world's marine ecosystems. Cambridge: Cambridge University Press.

International Ocean Institute, 2020. The future of ocean governance and capacity development. Leiden: Brill.

Kaiser, M., Attrill, M., et al., 2011. Marine ecology: Processes, systems, and impacts. 2nd ed. Oxford: Oxford University Press.

Kim, E.S., 2013. Marine renewable energy handbook. London: Routledge.

Klein, N., 2011. Maritime security and the law of the sea.

Oxford: Oxford University Press.

Lane, T., 2001. The merchant seamen's war. Manchester: Manchester University Press.

McKibben, B., 2010. Eaarth: Making a life on a tough new planet. New York: Times Books.

Nicholls, R.J., Wong, P.P., Burkett, V.R., et al., 2007. Coastal systems and low-lying areas: Climate change impacts. Cambridge: Cambridge University Press.

Paine, L., 2013. The sea and civilisation: A maritime history of the world. New York: Knopf.

Pauli, G., 2010. The blue economy: 10 years, 100 innovations, 100 million jobs. Taos: Paradigm Publications.

315

Sampson, H., 2013. International seafarers and transnationalism in the twenty-first century. Manchester: Manchester University Press.

Sorensen, H.C., 2011. Ocean energy systems: Overview of ocean energy technologies. 3rd ed. Paris: International Energy Agency.

Stopford, M., 2009. Maritime economics. 3rd ed. London: Routledge.

Symonds, C.L., 2001. Decision at sea: Five naval battles that shaped American history. Oxford: Oxford University Press.

Talley, W.K., 2011. Maritime safety, security and piracy.

London: Informa Law.

Thorpe, T.W., 1999. A brief review of wave energy. London: Department of Trade and Industry.

Till, G., 2013. Seapower: A guide for the twenty-first century.

3rd ed. London: Routledge.

Urbina, I., 2019. The outlaw ocean: Journeys across the last untamed frontier. New York: Knopf.

Vega, L.A., 1999. Ocean thermal energy conversion (OTEC).

2nd ed. New York: Springer.

About the Author

Artur Victoria is a Researcher at OBSERVARE – Observatory of Foreign Relations of the Autonomous University of Lisbon (UAL) and lecturer in non-governmental organizations,

Graduated in Law in the University of Lisbon. Degree in Defense Course and holds certificates such as: Trainer of Trainers (BAR Association, update course at the National Defense Institute, Diploma on Corruption Control (Open Society Hungary) .He was the founder of the Colegio Luso Internacional - CLIP in Portugal, President of the European representative of the Portuguese Culture Federation, Founder and President of the Brazilian – Portuguese Law Institute - , Representative in Europe of the Association of Graduates of the Escola de Diplomados Superior de Guerra - Brazil. He practiced law for 21 years - he was an elected member of the District Council of the Portuguese Bar Association of Lawyers and a

trainee advisor at the BAR Association.

Author of five law books, he writes articles on ethics and security for academic websites.

Official commendations:

- 2014 Brazil - Friend of the Navy Medal (Brazilian Navy)

2015 - Portugal- Medal of Merit of the Portuguese Army D.Afonso Henriques - silver grade

2015 - Brazil- Order of Merit Aeronautico (Knight) of the Brazilian Air Force

2019 Brazil - Medalha Tamandare (Brazilian Navy)

2020- Brazil - Order of Naval Merit – Knight (Brazilian Navy)

2021 - Portugal - Medal of the Cross of St. George 1st degree 2020 - Brazil -Ordem do Merito promoted to Officer (Brazilian

Air Force)

2021 Brazil - Order of Naval Merit – (Brazilian Navy)

2023 Brazil-Order de Rio Branco – given by the President of Republic of Brazil

Read more at https://observare.autonoma.pt/investigador/artur-victoria/.

About the Author

Artur Victoria is a Researcher at OBSERVARE – Observatory of Foreign Relations of the Autonomous University of Lisbon (UAL) and lecturer in non-governmental organizations,

Graduated in Law in the University of Lisbon. Degree in Defense Course and holds certificates such as: Trainer of Trainers (BAR Association, update course at the National Defense Institute, Diploma on Corruption Control (Open Society Hungary) .He was the founder of the Colegio Luso Internacional - CLIP in Portugal, President of the European representative of the Portuguese Culture Federation, Founder and President of the Brazilian – Portuguese Law Institute - , Representative in Europe of the Association of Graduates of the Escola de Diplomados Superior de Guerra - Brazil.

He practiced law for 21 years - he was an elected member of the District Council of the Portuguese Bar Association of Lawyers and a trainee advisor at the BAR Association.

Author of five law books, he writes articles on ethics and security for academic websites.

Official commendations:

- 2014 Brazil - Friend of the Navy Medal (Brazilian Navy)

2015 - Portugal- Medal of Merit of the Portuguese Army D.Afonso Henriques - silver grade

2015 - Brazil- Order of Merit Aeronautico (Knight) of the Brazilian Air Force

2019 Brazil - Medalha Tamandare (Brazilian Navy)

2020- Brazil - Order of Naval Merit – Knight (Brazilian Navy)

2021 - Portugal - Medal of the Cross of St. George 1st degree

2020 - Brazil -Ordem do Merito promoted to Officer (Brazilian Air Force)

2021 Brazil - Order of Naval Merit – (Brazilian Navy)

2023 Brazil-Order de Rio Branco – given by the President of Republic of Brazil

Read more at https://observare.autonoma.pt/investigador/artur-victoria/.